D0629418

FOREWORD BY JUDITH MILLER
FOREWORD BY ALAN P. ZELICOFF

INSIDE THE
SOVIET/RUSSIAN
BIOLOGICAL
WAR MACHINE

IGOR V. DOMARADSKIJ
& WENDY ORENT

Prometheus Books
59 John Glenn Drive
Amherst, New York 14228-2197

Published 2003 by Prometheus Books

Inquiries should be addressed to
Prometheus Books
59 John Glenn Drive
Amherst, New York 14228–2197
VOICE: 716–691–0133, ext. 207
FAX: 716–564–2711
WWW.PROMETHEUSBOOKS.COM

07 06 05 04 03   5 4 3 2 1

Library of Congress Cataloging-in-Publication Data

Domaradskiæi, I. V. (Igor§ Valerianovich)
    Biowarrior : inside the Soviet/Russian biological war machine / by Igor V. Domaradskij ; with Wendy Orent.
        p. cm.
    Includes bibliographical references and index.
    ISBN 1–59102–093–X (cloth : alk. paper)
    1. Biological warfare—Soviet Union. 2. Domaradskiæi, I. V. (Igor§ Valerianovich) I. Orent, Wendy, 1951– II. Title.
UG447.8.D65 2003
358'.38'0947—dc21

2003010856

Printed in Canada on acid-free paper

To our children

Say not that life is a toy
In the hands of unthinking fate,
A feast of idle inanity
And the poison of doubt and strife.
No, life is the reasoned aspiration
To where the eternal light is burning,
Where man, the crown of creation,
Rules the world from on high.

S. Y. Nadson (Russian poet, 1862–1887)

CONTENTS

FOREWORD

JUDITH MILLER

The breathtaking ambition and scale of the former Soviet Union's biological warfare program are now well known. The Soviet effort was nothing if not "bolshoi"—or "big"—the Soviet Union's supreme compliment.

For over fifty years until the Soviet Union's collapse in 1991, as many as forty to sixty thousand Soviet scientists, doctors, engineers, and technicians worked to make deadly germs ever more so in literally dozens of secret weapons facilities throughout the vast former empire. Many of these labs, institutes, and even secret cities did not appear on official maps; most were identified only by anonymous post office box numbers.

Much about the program remains unknown. Among the most disturbing anomalies is the virtual absence of the voices of its participants. Almost no former scientists have written about the program, or their part in it. The exceptions are mainly defectors—those who fled both their jobs and their country.

The most important book in English is *Biohazard*, the extraordinary memoir of Ken Alibek, a Soviet of Kazakh origin who rose to the second highest post in Biopreparat, the civilian cover for part of the bioweapons program. But Dr. Alibek fled the Soviet Union in 1992 as it became clear that neither the beleaguered union nor its high-priority weapons program, then at its peak, would survive.

That is one of the many reasons why Igor Domaradskij's book is so welcome. Domaradskij did not leave Russia, and his book was first published in Russian—a relatively daring act of quiet rebellion at the time. His is the voice of a man who witnessed the development of the Soviet biowarfare program from a unique perch almost since its inception. As the head of one of his country's leading antiplague institutes, which also eventually become part of the bioweapons infrastructure, he chronicles his transformation from scientist to at first unconscious, and then witting and willing biowarrior. He describes his own part in the creation and direction of Biopreparat, the supposedly civilian organization which served as a cover for the Soviet Union's military effort to create new and improved germ weapons.

He grieves for Russian science, then and now. Anyone who has ever spoken to a former Soviet biowarrior will understand his fury over the extent to which ideology, mismanagement, greed, ambition, and the militarization of research warped Russia's science. Those who have never even met a Russian may find his tales of crippling secrecy, bureaucratic turf wars, and personal vendettas exaggerations. But to the small group of biowarfare experts, they ring true, all the more so because Dr. Domaradskij acknowledges the extent to which he "spoiled his relationships" over time with powerful people who could have saved him from his more self-destructive impulses to take on a system that could not be changed in the name of Russian science.

He remains skeptical not only of fellow Russians, but of westerners too, and in particular, of Americans, the cold war's victors. Even now, he writes, he finds it "difficult to believe" that the United States abandoned germ weapons in 1969 as it claimed, a skepticism shared by some Americans as well. And he is equally leery of American and international programs that offer former biowarriors research grants and aid to help them through these economically troubled times. He fears that such efforts often mask unprecedented espionage efforts.

Without much self-pity, he clearly feels that he was co-opted professionally. And although he held senior posts in several different parts of the germ enterprise, he claims to know little about parts of the program in which he did not work. This, too, rings true, for the Soviet germ warfare program was shrouded in "legends," plausible civilian covers for military research and development, and heavily compartmentalized.

He deeply resents the fact that he was not permitted to travel or publish what he claims were important discoveries—such as the development of a new strain of hemolytic plague—in scientific journals. He shows the misuse of precious resources by scientists and institute directors who scrambled to seize whatever they could for themselves and their labs. He deplores the "mania for secrecy which was pursued to the point of absurdity."

He asserts that despite all the money and effort Moscow spent, the Soviet germ warfare program failed to achieve many of its goals, though it is unclear whether he was in a position to know what was accomplished by parts of the program in which he did not work.

His is not a politically or morally simple tale. He expresses no particular regard for the very few Russian scientists who quietly refused to turn genetic plowshares into swords. The strength of his narrative, in fact, stems from his candor about his own complex motives for standing science on its head in pursuit of personal and scientific advancement, and what he calls "patriotism." At first he tried to delude himself into thinking that the effort to develop new, more virulent strains was separate from the one to weaponize such agents. But eventually, he comes to see that the development of strains of microorganisms with specific properties, the field in which he worked, was but "a prelude to the real business."

Though much of the civilized world disagrees, he contends that germ weapons are not inherently more inhumane than guns, tanks, bombs, and other weapons that kill people. But his return again and again to the moral implications of his work belies that facile justification.

He freely admits that he was no "refusnik,"and he remains inherently skeptical of many of Russia's new reformers, who "having donned the robes of democracy" persist in "their illusion in order to stay in power." He remains equally scathing about those who seek modern Russia's salvation in the nation's traditional Orthodox Christianity. "Their newfound religion merely demonstrates a complete lack of principle," he writes. And he worries deeply—as should non-Russians—about the fact that many of the former Soviet officials and military officers who helped lead the bioweapons program still hold positions of significant influence.

He acknowledges that although he felt an "urgent need" to write about the purges that crippled early Russian science, the dramatic repression and KGB murders—even among his own family—there is much that he still cannot say. But this wry, candid memoir of an ambitious, obstinate "inconvenient man," the title of his book in Russian, goes further than most other accounts to illuminate an endeavor that many Russian officials continue to deny to this day.

He also worries about the fate of Russian science in this difficult transition from Communism. His work will undoubtedly be much appreciated by all those seeking to understand the former Soviet Union, and the price its citizens paid for their profound self-delusions.

FOREWORD

ALAN P. ZELICOFF
SANDIA NATIONAL LABORATORIES

The biological weapons program of the old Soviet Union encompassed a vast archipelago of facilities: laboratories for basic research, "antiplague" institutes where clinical experimentation took place, outdoor and enclosed testing sites for exposing animals to a wide variety of organisms and toxins, and production centers—geared to the manufacture and stockpiling of hundreds (perhaps thousands) of tons of microbes for use as weapons.

At its peak in the early 1980s, as many as one hundred thousand individuals—scientists, clinicians, administrators, and military personnel—were directly involved in this enterprise. Some of these were physicians and researchers internationally known for humanitarian work: V. M. Zhdanov, who first proposed the worldwide smallpox eradication campaign, also held a high position in the Soviet biological weapons (BW) program. P. N. Burgasov, chief medical officer of the Soviet Union and deputy minister of health, worked for the Soviet military BW program, and deflected international inquiry into an outbreak of pulmonary anthrax in the city of Sverdlovsk in 1979, explaining it all away to gullible investigators as contaminated meat—a position he still unaccountably maintains. Gen. E. I. Smirnov, head of the Soviet military's "15th Directorate" for many years—

the cover for the entire illegal biological weapons program through the collapse of the Soviet Union—led the Soviet (and later Russian) delegation to the Biological Weapons Convention. There are many others.

Slowly, a picture of the Soviet BW program—though still incomplete—has begun to emerge, thanks to the revelations of Vladimir Pasechnik, former head of the Institute for Ultra-Pure Preparations in Leningrad; Ken Alibek (K. Alibekov), Deputy Director of "Biopreparat," the shell organization that hid much of the BW program, and now Igor Domaradskij—physician, administrator, and one of the bioweapons designers in the early days of the revived Soviet program.

All of these accounts fill in much that was obscure in our understanding of the Soviet BW program. We can now say with some confidence that, after the Russian civil war following the November 1917 revolution, a fledgling biological weapons program began under the tutelage of A. N. Ginsberg, director of the Scientific Institute of Health in Moscow. Noting that infectious disease was the principal cause of mortality and suffering during the civil war, Ginsberg reported to defense kommissar K. E. Vorisholov on the feasibility of use of microbes as weapons, and also on the need for defense against such weaponry. Despite Soviet accession to the 1925 Geneva "Protocol for the Prohibition of the Use in War of Asphyxiating, Poisonous or Other Gases, and of Bacteriological Methods of Warfare,"[1] a research program into the development of biological weapons was initiated by decree from above in 1928.

Accounts vary on the details of how this decree was implemented, but it is clear from Domaradskij's account that anthrax, plague, cholera, and tularemia organisms were isolated, grown in mass, and tested on animals in several laboratories. By the late mid-1930s, open-air testing was begun on Renaissance Island (Vozrozhdeniye Ostrov), a tiny island in the Aral Sea. L. P. Beria, Stalin's minister of internal security, took special interest in the work, and by 1939 he was put in charge of the entire BW program, all of which was buried in various departments of the Ministry of Defense.

Almost nothing has been known about the Soviet program in the 1950s and 1960s; fortunately, Domaradskij's memoir begins to fill in that history. It seems that once the Soviet military acquired nuclear weapons in 1949, interest in BW waned. But Col. Gen. E. I. Smirnov, minister of health of the Soviet Union from 1947 to 1953 and then head of the 15th Directorate till his death in 1985, kept the program alive largely (though not exclusively) in a net of the military facilities. Rumors of a secret study of the World War II experiments of the infamous Japanese Unit 731, use of biological weapons by U.S. troops in Korea, and perhaps fear of Chinese expansionism all helped to drive a resurgence in the Soviet BW program. But little substantive progress had been made for twenty years. Domaradskij tells us why: scientific advances in biology—including in the development of biological weapons—had been stymied by a Soviet version of political correctness known as Lysenkoism, after the infamous Stalin sycophant T. D. Lysenko. Lysenko resurrected the long-disproved eighteenth-century notion that traits acquired during life would be passed on to subsequent generations, rejecting even the existence of genes in a self-serving attempt to aggrandize Stalin's policies. After all, realizing a communist utopia (and timeless adulation of the great leader) merely required adherence to a strict Stalinist philosophy that would then be acquired by the next generation at birth. Beria ruthlessly enforced Stalin's Lamarkian vision; scientists who objected were executed or sent away to the camps.

Thus the military BW program stagnated, even after Stalin's death. Its rebirth in the early seventies depended on the work of scientists like Zhdanov and Domaradskij, both of whom did much to put the Soviet bioweapons program on a firm footing rooted in a knowledge of modern molecular biology and genetics.

Ironically, at the same time that it developed a modern BW program, the Soviet system—as well as Domaradskij himself—devoted considerable effort to public health. Bolshevik revolutionaries realized decades earlier that infectious disease—typhus in particular, which was spread by lice—was exacting an enormous toll on people in cities and the countryside, let alone the Revolutionary army itself. Lenin famously noted: "Either socialism will defeat the louse or the louse will defeat socialism." A string of "antiplague institutes" was created and scattered in a wide swath across the new Soviet Union. They survive today as part of the

Russian public-health system and were quickly seen as relatively apolitical outposts of scientific excellence.[2]

Military officials kept close watch on the work of people like Domaradskij, who was appointed director of the antiplague institute in Rostov-on-Don in 1964, after many years of productive work at an antiplague institute in Irkutsk. There he battled a cholera epidemic and found creative ways to make up for the lack of funds to outfit a laboratory; he even implemented his own culture media production line so that organisms could be isolated and identified. He also managed to find time to write monographs and to work in the local Soviet bureaucracy. During his tenure in Rostov, Domaradskij first became aware of "Problem No. 5"—the secret Soviet biodefense program, which later became a cloak for the still more secret "Problem Ferment," the code name for the entire Soviet BW program. "Ferment's" activities, for the most part, were carried out in the Biopreparat system; the institutes of the Ministry of Health were also involved, though to a lesser degree. Created as a nominally civilian system by the Military Industrial Commission under the USSR Council of Ministers, this civilian BW infrastructure under the Main Directorate of the Microbiological Industry (Glavmikrobioprom) was designed to supplement the flagging military efforts. The name "Biopreparat" became well known in the West only after Alibekov told the world about it after his 1991 defection to the United States.

Domaradskij's successes in Irkutsk and Rostov-on-Don did not go unnoticed. Powerful military and civilian leaders, eager to have post-Lysenko science reintroduced into the biological weapons program, recognized Domaradskij's organizational and scientific talents. Thus, in the early 1970s Domaradskij was recruited to hold an important administrative position, one which gave him a rare vantage point from which to view almost all of Biopreparat's scientific activities, though much of what happened at the military labs remained outside his purview. Even today, as Domaradskij makes clear, the activities that take place at military sites such as the Virology Institute (in Sergiev Posad just north of Moscow), the Research Institute of Epidemiology and Hygiene (in Kirov 600 miles to the east of Moscow), and "Compound 19" in Ekaterinburg (formerly Sverdlovsk)—the origin of the accidental release of anthrax and subsequent epidemic among animals and humans in 1979—are cloaked in darkness.[3]

Once in Moscow, Domaradskij was thrust into the world of the professional apparatchik, struggling with the countless egos and personal agendas of the ruling elite. This memoir, expertly rewritten with Wendy Orent, is the chronicle of the good years and bad, and his attempt to reconcile his patriotism (and his work developing ever more deadly weapons for a bureaucracy he despised) with his love of science and knowledge.

Could there still be an offensive biological weapons program in Russia?[4] We simply don't know. A Russian proverb says: "All cats are gray at night." We still don't know what color to paint the remnants of the Soviet biological weapons program, but Domaradskij's memoir sheds some new light that may enlighten our understanding and lead, eventually, to better cooperation. Stalin liked to describe the (inevitable) transition from capitalism to communism as a methodical and sometimes painful process[5]—"step-by-step" he wrote—throwing out the bourgeoisie and replacing them with the ideal Soviet citizen. We seem destined to learn about the intent, scope, and technology behind Problem Ferment in just the same way.

NOTES

1. The Geneva Protocol did not ban the acquisition of biological or chemical weapons, merely their use. The United States did not accede to the protocol until 1975 but President Richard Nixon unilaterally renounced the stockpiling of biological materials for weapons purposes and all research into offensive use of biological weapons in 1969.

2. The antiplague institutes were extremely successful, if somewhat heavy-handed, in eliminating much vector-borne disease in the Soviet Union.

3. As recently as August 2002, Sen. Richard Lugar was denied access to Kirov on a visit to nonconventional weapons sites in Russia; ironically, Lugar is the political and intellectual champion of the Nunn-Lugar nonproliferation initiative, by far the largest source of hard currency for the Russian biological

weapons labs (although how much of it winds up in the military labs is unknown). Over a billion dollars has gone to Russia to secure and eliminate weapons of mass destruction (nuclear, chemical, and biological) over the past decade under the Nunn-Lugar legislation.

4. The Soviet Union became a signatory and co-depository (with the United States and Great Britain) to the Biological and Toxin Weapons Convention (BTWC) in 1972. The BTWC entered into force in 1975 and prohibits research into offensive biological weapons or stockpiling of biological materials in types or quantities that have no justification for peaceful purposes. Russia assumed the responsibilities of this treaty in 1992. The Soviet Union was never investigated for violation of the BTWC after the Sverdlovsk anthrax event of 1979. The BTWC has no explicit verification provisions, but does provide for cooperation among states parties to resolve concerns.

5. See, for example, J. V. Stalin, *The Foundations of Leninism*, lectures delivered at the Sverdlov University and published in *Pravda*, nos. 96, 97, 103, 105, 107, 108, 111; April 26 and 30, May 9, 11, 14, 15, and 18, 1924.

COAUTHOR'S NOTE

I gor V. Domaradskij, plague fighter and plague designer, has had what must surely be one of the most remarkable careers in microbiology in the later twentieth century. In a society not noted for its meritocracy, he rose from an obscure and—from the perspective of the Soviet system—politically suspect background to become one of the chief architects of the entire secret biological weapons program as well as one of its principal scientists. The Russian-language version of this memoir, titled *Perevertisch*, which may be roughly translated as *one who says one thing and means another*, caused him no little trouble with the FSB, the Russian descendant of the infamous KGB. For five years after its private publication in 1995, he lost all right to travel outside the country. A copy of this book fell into the hands of the U.S. Department of Defense, where it was translated into English by an anonymous translator. The English-language manuscript, called *Troublemaker*, circulated samizdat-fashion among biodefense experts, researchers, and journalists in the United States and Britain.

That manuscript—with thanks to the original translator—formed the basis for the present publication, which has been extensively rewritten and expanded. But the tone of the original text is preserved. The Western reader may find that this book has more the flavor of a Russian novel than of a modern memoir. Domaradskij is reticent about his private life, and even more so about his actual position in the bioweapons hierarchy, where he initially played one of the principal roles in designing the entire program. Ken Alibek,* the well-known Russian defector, former second-in-command at Biopreparat, and author of the best-selling memoir *Biohazard*, once told me that Domaradskij was like a god in the eyes of the younger scientists. This does not come across in the text, as the reader will see. Domaradskij is a modest man, and does not think to assert his own importance.

The perceptive reader will also notice, however, an unmistakable scent of score-settling at certain places in the text. Domaradskij was pushed out of the biowarrior program by administrators he despised, and he is telling his side of the story. History backs him up, however; the notorious mismanagement of the Institute of Applied Microbiology, located in Obolensk, has resulted in the total breakdown of the facility—extending even to the threatened cut-off of electrical power—and the consequent risk to the public from dangerous pathogens that are no longer adequately maintained and safeguarded. Domaradskij's bête noire, Gen. Nickolay Urakov, director of the institute, unregenerate Lysenkoist and foe of modern genetics, has recently been fired for alleged mismanagement. General Kalinin, too, has been removed from power.

Despite the harshness of some of Domaradskij's observations, they have been preserved intact. No attempt has been made to neutralize or soften any of his views—they are his own. Furthermore, as his coauthor, I have labored to preserve the essential foreignness—the Russianness—of the text. I view my role as one of making the book comprehensible without any unnecessary distortion of his mode of thought and expression. I did not wish to prettify the text, or Americanize it, or transform it into something foreign to the person whose story it is. This was a difficult task, since Domaradskij's English is entirely self-taught and was therefore in need of extensive revision to render it comprehensible. Working together through e-mail, Domaradskij and I produced several drafts, each

---

*Kanatjan Alibekov.

draft adding more and more exposition and explanatory material. For the final two revisions of the book, we had the good fortune to work with Steven L. Mitchell, editor-in-chief of Prometheus Books, who e-mailed us extensive commentary and queries, forcing us to clarify, expand, and explicate much of the narrative. The result is a book almost twice as long as the original *Troublemaker* text.

Here is the man himself, in his own words, as much as possible, for history to judge as it will.

ACKNOWLEDGMENTS

S pecial thanks go to Marina Nevskaya for her support of the original manuscript. Without her help the Russian edition would never have been published. I also want to thank Wendy Orent, my coauthor for the English-language edition. Hers was a huge job transforming and refining the work for the English-language audience. Her excellent writing skills have improved the text tremendously. I would also like to thank Julian Perry Robinson for his valuable advice as we prepared this edition. To Judy Miller of the *New York Times*, who was one of the first Westerners to read the Russian edition, I owe a great debt of gratitude. Her unwavering support was instrumental in securing a publisher for this work. Appreciation is also due to Ray Zilinskas for his interest in the theme of the work, as well as to Donald Atlas, who published an early version of a portion of the text in the journal *Critical Reviews—Microbiology*. The book is further enriched by the explanatory notes contributed by biological and chemical weapons expert Dr. Benjamin C. Garrett of the Federal Bureau of Investigation. I thank him for his valuable additions to the text.

I am indebted to my wife, my family, and my friend Vladislav Yankulin for their patience and support during a very difficult time of my life.

PROLOGUE

Until now I had never written anything but scientific articles and books. I therefore beg your indulgence for this attempt to tell you about my affairs, which at times are purely personal. I'm afraid this may seem strange to some and tedious and uninteresting to others. But as Leo Tolstoy once said in a different context, "I cannot be silent!" I am now of an age at which the urgent need has arisen to speak out, to examine my own actions and my life. I feel an obligation to do so because of the tremendous upheavals of the recent past, and also because of our present unstable situation—the beginning of a new life after the collapse of the Soviet empire and the making of a new democratic society in Russia—for which we are all answerable.

I flatter myself with the hope that the testament of a man who has lived through the years of Stalinist repression, hunger and war, the nightmare of various false doctrines, the "Khrushchev thaw," the years of stagnation, the "restructuring,"* and the collapse of the Soviet empire will be an encouragement to someone who feels impelled to write on the same subject, only better than I, as I am a scientist and not a writer.

---

*The author is referring to the process widely known as *perestroika*, which might be translated as "restructuring" or "reworking."

I have tried to set out the events of my life in chronological order, although I have sometimes had to digress or run ahead if this makes for a better examination of some particular facts and events. I have also tried to be as accurate as possible in describing these facts and events, avoiding any exaggeration. However, after this span of time, and because I could not always obtain complete information, I may not have been able to avoid some mistakes in this account. Each man commits many blunders during his life, as all of us are human, and none completely innocent. For any such unintended errors I apologize. As for the interpretation of events, the assessment of certain individuals and their actions, these are mostly of a personal nature and are a reflection of the circumstances in which I found myself.

For this book I have been mainly indebted to my wife, a unique personality and an excellent actress who gave up her beloved career for my sake. I often wonder whether I was entitled to accept this sacrifice and whether I have done enough to repay her. At the same time I also feel bound to express my deep gratitude to those of my colleagues without whose goodwill and assistance, material as well as moral, this story would hardly have seen the light of day. I hope they will forgive me for not mentioning them by name. I do not wish to endanger them for helping me write this "seditious" book. Given the state of things as they still are in my unsettled country, I cannot even now say all I know: "God spares the discreet."

CHAPTER 1

O tempora, O mores!

(Cicero)

his is the story of an "inconvenient man"—a "troublemaker" as I have learned to call myself. The son and grandson of men persecuted by the Stalinist system, I learned to work within that system to survive. I am a scientist, and much of that which has driven me in my life has been the desire to "do science." To achieve that goal, I had to work within the very system that had poisoned my early years and killed many members of my family. I grew up in an atmosphere of cruelty and violence, which directly touched some of the people (and not only near relatives) whom I loved most, and which changed radically the direction of our lives. I could never forget that; I could never become reconciled to the surrounding reality of the Soviet political system. But despite the loss of so much, and of so many loved ones, I found that life went on. To survive, I had to hide my true attitude toward the Soviet regime from childhood; I learned quite early to adapt myself to this regime. Such an adaptation was a necessity, which came down in essence to the Shakespearian question: "To be or not to be?"

To understand how I came to be a troublemaker,[1] to examine my character, and to explore the reasons for the many adventures I have experienced, before anything else I need to describe my ancestry. In spite of the long litany of hardship and persecution that befell my people, I am proud of my family and have never forgotten them.

My great-grandfather on my mother's side, Ivan Yakovlevich Drevitskij, was of peasant stock from the Bakhmach district in the province of Ekaterinoslav (present-day Dnepropetrovsk). He had the good fortune to be one of the original discoverers of coal and rock salt in what is now the Donbass. It made him prosperous. According to documents held in the Rostov Oblast[2] archives and to KGB records, by the end of the last century he was considered a member of the merchant class, having become a merchant in the first guild, and thereafter an honorary hereditary citizen, one of the privileged classes in old Russia. He owned a steamship line in Taganrog, a few miles west of the city Rostov-on-Don (one of his six steamers was still working the Azov and Black Seas until at least the 1960s), coal mines near Mushketovo, shares in the salt mines and other securities, a property outside Yuzovka (a town now called Donetz), a wharf in Rostov-on-Don, and a house in Taganrog (still standing to this day, at no. 3 Nikolaevskaya Street). His estate was assessed at about 800,000 rubles, a sum that in today's currency may translate into tens of millions of dollars. He was a member of the Taganrog municipal council and the mercantile exchange. Ivan Yakovlevich died comparatively young in 1912. The details of his will are unknown, so that no documents can be found as to the division of his property.

My great-grandmother's name was Ksenia Mikhajlovna, but I don't know where she came from. I remember my grandfather telling me that she was descended from some impoverished nobility.

My grandfather, Yakov Ivanovich,[3] was the eldest of their children, born in May 1881. Of his three brothers two died in youth and only one, Sergei, grew to maturity. There were also three sisters. One of them, Mariya, died in 1964. I never knew the others, and know very little about them. For some reason my grandfather was sent to study at the high school in Mariupol on the Azov sea, although his brother and sisters went to school in Taganrog. I know very little of my grandfather's early life. To judge from the scraps of information I have, my grandfather was living with another family, whose apparent lack of interest in his upbringing was bound to affect his chances of learning appropriate behavior and progressing in his studies. Nevertheless, he completed high school. In a report written on the conclusion of his studies the headmaster described him as "a modest and quiet young man, not lacking in ability; his natural character displayed some talent." Unfortunately, he did not do well at first; he passed through the junior and middle school in company that was rather deleterious to his moral education and growth outside of the classroom. By Form 8 (the American equivalent would be twelfth grade, the last year of high school) he was held back for a second try. "This was a salutary lesson" notes the headmaster's report: "This young man has now altered suddenly and significantly for the better, he has begun to work seriously and he has made good headway. If he keeps this up he could score some excellent results. He has always been polite to his elders, though he occasionally shows signs of stubbornness and willfulness. He is friendly toward his companions."

After high school my grandfather went off to Warsaw, Poland, to the Polytechnic Institute. Apparently he was not happy there; he left Warsaw very soon afterward. In 1901 he entered the Department of Physics and Mathematics at the University in Odessa, a large, cosmopolitan city along the Ukrainian shore of the Black Sea. There he took part in student political gatherings in the restless years between 1901 and 1905, which culminated in the failed 1905 Revolution.[4]

In August 1905 my grandfather applied to the principal for a transfer to the University of St. Petersburg. He gave family circumstances as the reason (his two sisters were studying courses at the Lesgaft Institute[5] in St. Petersburg, and they could not live there on their own—it wasn't considered socially acceptable). Permission was granted, and my grandfather continued his education in the natural sciences section of the Department of Physics and Mathematics at the University of St. Petersburg, from which he graduated in 1907.

While still studying at the Mariupol high school my grandfather came to know my grandmother, Sofiya Mikhajlovna (a very pretty girl!), who came from a modest, large Jewish family, and he married her in 1902.

My grandfather's family regarded this marriage as a bad choice. At all events I never saw my grandmother with my grandfather's family. I imagine that for the most part this attitude toward my grandmother was due not so much to her modest background as to the fact that she was Jewish, although she had to be baptized in the Russian Orthodox Church in order to marry my grandfather.

My mother, Zinaida, was born in Odessa on March 5, 1903, while my grandfather was still a student at the University of Odessa, and in 1907 my Aunt Oksana, her sister, was born, perhaps in St. Petersburg, although I am not quite certain.

Where my grandfather and his family went after his graduation, where they lived, and what they did I am not certain. To judge from a scrapbook my mother made in her girlhood and kept all her life—it is quite schoolgirlish, with some dedicatory verses and drawings—after 1910 the family lived in Moscow where my grandfather's sisters, all married, had already moved. There in Moscow my grandfather had a substantial house which is still standing (at 10 Neopalimovskij Lane). Here he became one of the founders of the Automobile Club-house (for car enthusiasts) and started to build a chemical factory. During the First World War he drove his own car from Moscow to serve as a volunteer at the front near Dvlnsk (a town in Latvia), where he repaired vehicles (as workshop foreman) until 1917.

By the time of the February Revolution in 1917 my grandfather's family had moved near Taganrog to a farm that belonged to my grandfather at Zakadychnayaj. My mother continued her studies there, but at home; she did not go to school. Her studies were directed by a governess named E. S. Ordynskaya, with whom we maintained friendly relations until her death in the late 1970s. I can't say whether it was due to Ordynskaya or not, but my mother was quite accomplished. She could speak German and French, and she played the piano quite nicely. She also knew how to draw.

To judge from my mother's account, after the October Revolution and the ensuing civil war between the Red Army and the rebellious Whites, she and her family had to travel about the Ukraine a great deal. It was quite a dangerous existence (owing to the forces of Nestor Makhno,[6] an anarchist who was one of the White civil war leaders in the Ukraine, and to the many bandits who overran the countryside). At some point, too, they lived in the town of Ekaterinoslavl. The Bolsheviks (the "Reds") fought the Ukrainian nationalists (the "Whites") to prevent Ukrainian independence. Eventually, of course, the Reds were victorious, and my grandfather, who had sought only to keep his family safe, was one of their victims. He had hoped, by these migrations, to get away from the Communists. But when he was arrested by the Bolshevik government for the first time, the police used his flights across the countryside as a pretext for accusing my grandfather of collaborating with Ukrainian nationalists.

In 1919 my mother's family left the Ukraine and returned to Moscow. There she finished her schooling in the early 1920s. In 1923 she married my father, Valerian Vladimirovich Domaradskij, who worked in a small office and who was fifteen years older than she. I never heard how my parents met and decided to marry, but this marriage was likewise greeted without enthusiasm, as my father was much older and poor.

I was born in December 1925. At that time my parents were living in a private flat in Starokonyushennyj Lane not far from the house in the Krivoarrbatskij Lane belonging to my grandfather's sister Mariya. At the age of one year I came down with poliomyelitis and lay in plaster for six months. Of course I remember nothing of this period, though immobilizing a child with a paralyzed limb seems to have been the standard treatment for paralytic polio in Russia during that era. I still limp with one leg because of the residual effects of the disease.

I know very little of my father, a taciturn man who never talked to me about his past. To learn the names of my paternal grandfather and grandmother, which I have done only recently, I had to turn to KGB records. (I have had to refer to KGB records quite a few times for details of my family history!)

My father was born in 1888 (or 1889) in Kielce, Poland. Where he came from and to whom he was related have remained mysteries to me. I do know that my father's family lived in Suvalki, Poland, on the frontier with what is now Lithuania, which was where my father graduated from high school. But before the outbreak of World War I, my father had already moved to Petrograd (as St. Petersburg was renamed after the Revolution) where he studied at the Higher Commercial Institute for a degree, which he earned in 1914. World War I broke the links with his parents; he did not return to Poland and never learned anything of their fate.

As for the identity of my paternal grandfather, Vladimir Fedorovich, I can merely quote the KGB records. These state that "Vladimir Fedorovich was employed as a bank cashier in Kielce and later as a magistrate in a town near Suvalki"—the records aren't very specific. Since my father's family apparently remained in Poland, we were completely cut off from them after the war and the Bolshevik Revolution. I know nothing for certain about my father's family history. But as a child, reading one of my parents' books, I found a surprising clue to the mystery of my ancestry. The writer D. Mordovtsev, in his well-known book *False Dmitri* about a pretender to the Russian throne after the death of Czar Boris Gudonov in the years 1605–1612, several times mentions one Pan Domoratskij, a noble Polish vassal of the Sandomir voevod,[7] Prince Yuri Mniszek. The name Domoratskij is quite uncommon: perhaps my father's family descended from this Polish noble. It may be that after the October Revolution my father, fearing persecution, russified our family name (this is confirmed by the records of a tribunal, an extract of which I quote below). I also know that in 1921–22 my father visited the Lithuanian mission in Moscow several times (after realignment of the frontier, Suvalki passed to Lithuania), applying for "permission to go there in order to look for his relatives, although this was withheld as he was the son of a Russian official."

My father was a devout man and I remember him often going to church with my mother. (Despite her churchgoing, my mother remained an atheist to the end of her days, just like her father.) Father was always distant with me, always reticent in conversation, and it was only after I had become a student that he showed any interest in me. He never told me anything about himself or his parents, and he left no papers behind. I kept only a tiny photograph which I later had enlarged. It shows my father in his office seated at an enormous typewriter when he was working in

Ekibastuz before the Revolution and where, according to his official title, he was "the secretary and manager in the general affairs department of the coal mines and factories of the Kirgiz Mining and Industrial Company."

I have never found any explanation for his attitude.

My father had no living relatives. I have been able to learn of only one great-uncle in Moscow, Mikhail Moiseevich Domaradskij, about whom all trace has been lost. I only learned about him from KGB records in the 1980s. But recently, by chance, I have found, through an Internet search, descendants of M. M. Domaradskij right here in Moscow. They are very pleasant people and we have become friends. They told me that in 1940 M. M. Domaradskij and his wife were arrested by the KGB, too, and both died in prison a year later—more victims of Stalin's terrorism!

My father had no friends either, and I never saw any guests arrive at our house. It was only occasionally on days off—in the 1930s we had "days off," Sunday having been abolished—that some near relative of my mother's might dine with us. We never celebrated revolutionary holidays at home.

In 1927 my father, maternal grandfather, his brother (my great-uncle) Sergey, and the husband of his sister Mariya were arrested. They were charged under articles 58-5, 58-10, and 58-18 of the criminal code[8] of the Soviet regime. Article 58 was the most terrible and serious of all charges (espionage, sabotage, anti-Soviet activity, and so on). It no longer exists in the new Russia. My father, unaccountably, was accused of spying for Lithuania! My grandfather was alleged "to have consorted with the Whites" (rebellious nationalists who fought against the victorious Bolshevik army after the October Revolution), and, in particular, with the Ukrainian hetman[9] Pavel Skoropadski, until the Whites were finally liquidated. The Cheka (later known as the GPU)[10] also accused him of having suspicious "connections abroad" with Lithuanians, and to have held and permitted meetings in his flat which involved "anti-Soviet attacks." After three months, all of those arrested were released, but by Decree of a Special Commission under the directorate of the GPU they were denied the right to live in a number of cities in the USSR: Moscow, Leningrad, Kiev, Rostov, Odessa, Kharkov, and any frontier districts for a period of three years (at that time the maximum period of expulsion).[11] Such exile was one of the many forms of repression invoked if, for some reason, the Cheka could not apply other, more savage sanctions in a particular instance. Exiled people were deprived of any civil rights including the rights to work and to move about freely.

My father and grandfather chose Saratov for their exile. There is a request on file from my relatives to delay departure from Moscow because of my aunt's illness, but it was denied. Thus from the age of two, and for most of the next thirty years, I lived in Saratov. Except for frequent trips, and for one year which I spent with my grandfather, I did not return to Moscow again until 1973, when I was nearly fifty.

At the same time, one of my grandfather's sisters, Alexandra, disappeared. Perhaps she died: her husband had been shot by the Reds during the civil war. Later her son, my mother's cousin Nikolai Gaponov, also disappeared; he had been arrested in 1927 together with the other relatives. After he was released from prison, he settled in Sverdlovsk together with his uncle Sergey Ivanovich Drevitskij, who had formerly served in the Volunteer Army—rebels against the Bolsheviks. Later it was consolidated as a regular army and known as the White Guards under Gen. A. Denikin.[12] Another of my grandfather's sisters, Anna, together with her close relatives, lived abroad, probably in Yugoslavia. Thus, out of this once large family there remained only my grandfather, his brother Sergey, and his sister Mariya—the only one of the sisters I remember.

For his place of banishment Mariya's husband, Ivan Alekseevich Sedin, a lieutenant colonel who had served in the Russo-Japanese War, chose Yaroslavl, his birthplace. Their daughter, Tatiana, who was a few years younger than my mother, in 1929 met and married a Hindu and went off to India. I saw them in Moscow twice in the mid-1930s. After that Tatiana was not allowed back into the country, although for many years she was an active member of the Society for Indo-Soviet Friendship. I once saw a photograph of her standing next to Khruschev at a reception for him in India. Tatiana returned to Moscow only once more when her mother, Mariya, died in 1964. She was late for the funeral.

My mother and my wife buried Mariya Ivanovna. As a member of the

Communist Party and director of the governmental Antiplague Institute in Rostov, I did not attend the funeral, as I feared to be seen meeting Tatiana. Connections abroad remained politically suspect: I have never admitted to having relatives living abroad on any job questionnaires, just as I have never written about politically suspect members in my family. In order to be hired for any job in the Soviet Union it was necessary to conceal these connections. Every time a person applied for a job or for university studies, he was asked to fill in such a questionnaire. He had to show who he was, essentially: what was his nationality or ethnic origin, his place of birth, his social status, his ties to the outside world, such as having relatives living in foreign countries, and so on. Having a relation like Tatiana abroad, therefore, could have been politically compromising for me.

Mariya Ivanovna's house in Krivoarbastskij Lane, which had somehow remained her property throughout the years of Soviet rule, was passed to the municipality (I think with some help from Tatiana) and has become part of the polyclinic of the 4th Main Directorate of the Soviet Ministry of Health. This is of particular interest as the whole of my grandfather's property was expropriated after the revolution, but after his banishment from Moscow the flat in Zemledelcheskij Lane was also lost (so far as I know, it fell into the hands of my grandfather's former household servant).

Fairly recently, from the KGB Administration for Moscow and the Moscow District, I received some photographs and various documents extracted from the files covering the arrests in 1927. Among them was a letter sent by my grandfather from Saratov to his brother in Sverdlovsk. In it he expressed his anxiety at the condition of Oksana, the youngest daughter, mortally ill with tuberculosis. How did this letter, which was sent from Saratov, end up with the GPU in Moscow? Were they tracking even my grandfather's correspondence in Saratov or that of his brother in Sverdlovsk? Whatever could they want it for? It is difficult for those who did not live through these times to understand the paranoia of the Soviet government, or the fear it instilled in us. To explain it now to those who did not experience it is almost impossible: even for us, it is difficult to remember the effects these fears and persecutions had upon us and our daily lives.

At first, life in Saratov was difficult for my family. As an exile, my grandfather was unemployed; he earned a living at various odd jobs, such as toy-making and composing puzzles (I have kept one to this day). My mother busied herself with making artificial flowers and took up embroidery (her "Asters" is still hanging in our flat). My father, a shorthand teacher by trade, organized the first private shorthand courses in Saratov in order to earn a modest living, which he continued to teach, together with my mother, right up to his death. At last, and quite by chance, my father managed in 1930 to get a flat. It was just two rooms, two rooms of 12 square meters each at 34 Nizhniya Street in an ordinary small Saratov timber house of one and a half stories. In the flat the only convenience was running water. Even at that time the place seemed tumbledown, and I never cease to wonder that it is still standing to this day!

The term of banishment ended in 1930 and my grandfather left for Moscow, where he took a job at the Compressed Gas Works. He also acquired a government-owned flat. His family had to stay behind in Saratov (for they no longer owned anything in Moscow). In the same year, in Saratov, another misfortune befell us. While trying to board a tram that had not come to a halt at the stop, my mother slipped off the platform and fractured her spine, leaving her an invalid for the rest of her life. My grandfather came immediately and took me to Moscow.

At first I stayed with some of my grandfather's relatives; later I lived with my maternal grandmother's sister Faina Mikhailovna Vogel in Solyanka Street. I had my fifth birthday there and was put into kindergarten. This kindergarten stood on the present site of the Military Academy in the rear premises of the Academy of Medical Science, of which I have now been a member for twenty-five years. As things have turned out, I am now working right next door, and every day I walk past the former kindergarten and the house that used to shelter me.

Because of my mother's ill health, my grandfather and I returned to Saratov a year later. My grandfather found a position as an inspector in the Department of Weights and Measures. Through the department he obtained a small flat with a kitchen, but with no other amenities whatsoever, though it had a splendid view over the Volga.

So far I have said nothing about my maternal grandmother: I don't know where she was at that time. She came with us to Saratov, and she buried her youngest daughter, Oksana, in Saratov. But what happened after that? For some reason she reverted to her maiden name of Rotenberg and kept it until her death in 1952. She may have wished to reduce the likelihood of persecution because of my grandfather's arrest. If so, it was extremely naive of her, since no one would have been likely to escape the attention of the GPU with such a ploy.

My grandmother was a child psychologist by profession. This science, along with eugenics and a few others, was at that time declared to be "reactionary" and "bourgeois." I have no idea why this was the prevailing government view. We cannot understand today many of the actions of the old Soviet regime. My grandmother, therefore, at the age of forty, had to be retrained. Where she graduated from training college, or how she made a living, I do not know. She turned up again in Saratov in the mid-1930s, some time after my grandfather returned. Up to her retirement at the beginning of World War II, she was teaching primary school classes in one of the prestigious grammar schools in Saratov and was considered a good teacher.

The Stalinist terror became stronger in 1937. Nearly every day we heard of someone being arrested. I cannot think of a family we knew in which somebody was not pulled in. They were locked up in droves. For example, the NKVD locked up our neighbor in the flat next door (he was in business somewhere), the father of a friend of mine (a barber), our housemaid's son (who worked on a trading post at a railway station), the husband of a relative of my mother, and a childhood friend who for some reason happened to be in Saratov (he was working at a combine factory that never produced a single combine!), and her daughter's husband (well known in Saratov as a master of ceremonies). Every knock on the door at night alarmed us. My father was afraid of everything. Perhaps this accounts for his secretiveness and strangeness.

Finally the terror reached us. In early November 1938 my grandfather was arrested again, once more under Article 58. This time it meant five years in the camps.

We heard later from the lawyer that the source of my grandfather's arrest was his deputy boss, one Zinovev. But the substance of the charge remained unknown. What most surprised us was that my grandfather got on excellently with this man and I was friendly with his nephews. So we were quite unable to believe in any treachery by Zinovev. Some years later and after my grandfather's death, during the war when we were really starving, Zinovev and his wife helped us by whatever means they could. He may have been trying to atone for his involuntary guilt (it is well known what means the NKVD used to get its evidence), or he may have been altogether blameless and acted out of the purest motives. When I was in my final year at the medical institute, I used to visit him in the hospital; he was suffering from cancer. Zinovev, and even more so his wife, were very interesting people, and I enjoyed their conversation enormously. They used to have plenty of books—a large library containing the works of many writers who were widely available only in the last years in the Soviet Union, but whom I got to know thanks to this couple. Their books were not what might be called antirevolutionary, but few people had access to classic literature in those days.

I remember my grandfather with his hands behind his back under escort by armed soldiers, the closed doors of the courtroom, the tedious wait for the sentence, and then the cold dark December morning, lining up to say good-bye at the prison before my grandfather was taken off to the camp in the town of Pugachev (not far from Saratov). I never saw him again. My grandfather suffered from tuberculosis, and in the camps it grew much worse. In one of his rare letters my grandfather asked to be sent books on medicine. We received the last letter from him when we were in Kzyl-Orda[13] at the end of December 1942. Then the chief of the camp sent us a telegram suggesting that we might come to collect him. Because of the war, my grandmother could not go and fetch him, but it would not have been of much use anyway; my grandfather died at the very beginning of January 1943.

As the flat occupied by my grandparents was officially owned, my grandmother was evicted immediately after Grandfather's arrest, and from then on she stayed with us.

My grandfather's brother Sergey Ivanovich also met a tragic fate. On May 31, 1944, by sentence of the military tribunal of the Sverdlovsk Railway, he was condemned for so-called anti-Soviet activities under Part 2 of Article 58-10 of the penal code of the RSFSR, subject to Article 58-2, to a ten-year deprivation of liberty in the corrective labor camps, with a three-year loss of civil rights and the confiscation of his property. In September that year he died of a "dystrophy" (from the records of the KGB Administration for the Sverdlovsk Oblast No. 16/0-342 dated June 23, 1988).[14] I met his wife several times later in Odessa. She was eking out a miserable half-starved existence together with her sister. In general, as the saying goes, "It never rains, but it pours."

In the 1980s when the Soviet perestroika—restructuring—had begun under Mikhail Gorbachev, I tackled the problem of rehabilitating the reputation and good name of my family members. This has taken several years and has produced a mountain of paper. I sometimes had to write several times to the same place, while letters were passed on from one office to another. Saratov was particularly distinguished for this sort of bureaucratic shuffle. I had no certainty of ever achieving any positive results, especially in matters dating back almost sixty years. But it seems that a piece of paper never goes missing in the files of our "guardians of justice"!

I began with my father. Rehabilitation means the restoration of rights, an avowal of innocence, and an admission that the original verdict was wrong. From the beginning of Gorbachev's "perestroika" in 1985, the matter of rehabilitation was decided by the Supreme Court, the military courts, or the Office of the Public Prosecutor. In my father's case, the matter was settled by the Order of Lenin Military Tribunal, Moscow Military District, on October 30, 1987. The wording of the order is typical: "The Decree dated 2 September 1927 In the matter of V. V. Domoratskij (Domaradskij) is hereby annulled and the case against him is quashed for want of evidence of any offense. Valerian Vladimirovich Domaradskij is hereby rehabilitated."

As for my grandfather, I attempted to rehabilitate him on the second charge, because it was the most recent. They would not consider it in Saratov. Instead, my application was passed to the prosecutor for the

RSFSR.[15] He entered an appeal from the sentence of the court with the Presidium of the Supreme Court of the RSFSR. By decision of the Supreme Court the sentence of the Saratov district court dated September 3, 1939, was annulled and the case was quashed for want of evidence of any offense.

The rehabilitation of Sergey Ivanovich was settled more easily: the decision to rehabilitate him was delivered by the district court in Sverdlovsk. In all these cases, I had to prove the ineptitude of the KGB actions, though that was never in doubt. I also tried to discover the whereabouts of my grandfather's grave. However, after lengthy correspondence I heard from Saratov that no records of this matter had been kept since "Corrective Labor Colony No. 7," where my grandfather had died, had been "disbanded."

I still wanted to learn at least something about my father's parents. It seemed to me that this might be done in Poland. Though I risked bringing trouble down on my head (because of my place of work we were forbidden to have any dealings with foreigners without official permission), I applied to the Consulate of the Polish Republic, who replied that "despite our careful inquiries we have failed to trace any descendants of your parents among the inhabitants of Suvalki." They asked me to solicit help from the editor of the newspaper *Przyazn* ("Friendship") in my quest. The paper published my appeal, which was answered by two people from Kielce and Suvalki. The result was discouraging, but I am grateful to them all the same. The woman in Suvalki even sent me a picture postcard of the town!

All of my inquiries are hindered by two factors. First, none of my close relatives are alive. Second, before she died my mother destroyed even the few papers she had, including a number of photographs she still kept. It seems that her fears and memories of the past had not left her even in her final minutes, although she died in 1980 when all (or almost all) of the terror lay behind us.

Of course, these terrors had an effect on my character, as did the compromises I made to survive within the system. But I never was a "good soldier"; I fought for what I believed in as a scientist. I always tried to be an honest man so far as possible within the restrictions of a conscience-less system.

NOTES

1. This was the English title of my original memoir, from which this book has developed.

2. Oblast in Russia is an administrative unit usually named after its main town (for example, Moscow—Moscow Oblast)

3. The author follows the Russian practice of referring to persons by their first name and patronymic, omitting the surname. The patronymic is taken from the father's first name. Thus Yakov Ivanovich would be the son of Ivan. The practice applies to females as well as males; thus Ivan's daughter would have the patronymic "Ivanovna." This practice is followed throughout Domaradskij's memoir. (Ben Garrett)

4. The government of Tsar Nicholas II imprisoned thousands and executed hundreds of people after the failed 1905 revolution. Leo Tolstoy issued a scathing criticism of this policy in 1908. That criticism was titled *I Cannot Be Silent* (in Russian: *Ya ne mogu molchat*). (Ben Garrett)

5. P. F. Lesgaft was a well-known Russian doctor and public figure. In 1905 he established the first institute for women's higher education.

6. The Western view is that Nestor Ivanovich Makhno (1889–1935) was a Ukrainian peasant anarchist who fought against both the Red forces loyal to the Communist cause and the White forces loyal to the monarchy or otherwise opposed to the Communists. Makhno and his supporters were defeated in 1921, after which he fled to France. Makhno was typical of Ukrainian peasants who fought against both Red and White forces during the Russian civil wars of 1918–1922 in the hopes of creating an independent Ukraine. Ukrainian nationalist sentiments were suppressed in Soviet times; hence, Makhno and his supporters were pictured as purely "anti-Red" or "pro-White" rather than as "pro-independent Ukraine." (Ben Garrett)

7. A *voevod* in Polish means a ruler of a province.

8. An accounting of the various criminal codes under which persons were charged and imprisoned during the Soviet regime is given in Alexandr Solzhenitsyn's *The Gulag Archipelago 1918–1956: An Experiment in Literary Investigation*. (Ben Garrett)

9. A *hetman* is a delegate ruler or commander.

10. GPU—Soviet political police agency, succeeded by the NKVD—The Public Commissariat of Internal Affairs—and later KGB—Committee of State Security. Now the agency is known as the FSB—Federal Service of Security.

11. One of the punishments meted out in the Soviet Union was to embargo an individual, denying him the right to obtain the permits needed to get a job or a place to live. Such "embargoed persons" were forced to find work outside the State system, placing them at risk of being arrested either for working without a permit and for "parasitism"—living off the State without providing anything in return. Such contradictions (being unable to work because one could not get a work permit yet being arrested for not having a job) were typical of life in the Soviet system, although they may be less familiar to Western readers. (Ben Garrett)

12. Recently I have written an essay on Gen. A. Denikin, "The Triumph and Tragedy of General Denikin," Historical newspaper (in Russian), March– April 1998.

Anton Ivanovich Denikin (1872–1947) was a career officer in the Tsarist army. He briefly served as commander-in-chief of Russia's Western Front in the summer of 1917, while Russia was still at war with Germany and Austro-Hungary, but was imprisoned by the Provisional Government for plotting its overthrow. He escaped following the November revolution and rose to command White forces loyal to the monarchy or otherwise opposed to the Communists. His march against Moscow failed, and the Reds defeated his forces at Orel. He resigned his command in March 1920, and fled to France. In 1945, he immigrated to the United States, dying in Ann Arbor, Michigan, in August 1947. His military record was mixed, and his command of White forces was unexceptional. However, he remained a popular figure with many Russian émigrés hopeful of restoring the monarchy. (Ben Garrett)

13. Where we relocated during the war: see chapter 2, p. 47.

14. In actuality, I think Sergey was shot. The phrase "dystrophy" or some such formulation was often used by NKVD to keep the knowledge from relatives that a prisoner had been executed.

15. RSFSR—Russian Soviet Federative Socialist Republic.

CHAPTER 2

GROWING UP

What is happiness? . . .
On your path through life
Wherever duty commands—go,
Know no enemy, measure no barrier,
Love, hope and trust.

A. Maikov (1821–1897)

n 1933 I went to school for the first time. I remember well the day
because it coincided with the end of rationing. Frightful hunger
had been raging along the Volga River (and in eastern Ukraine) as
a result of collectivization, a disastrous agricultural policy of the Stalinist
regime. Collectivization began in December 1929. Private farms, bee-
keeping associations, and all kinds of artels[1] were liquidated and were
transferred to the possession of *kolkhoz* or "collective farms."[2] All private
property, including cattle, horses, and any means of production were
passed to collectives. Millions of prosperous peasants (*kulaks*) were ban-
ished to Siberia, killed, or put into prisons and concentration camps for
three years. Thus, collectivization was accompanied by the wholesale
destruction of the peasantry. As a consequence, hunger was widespread
throughout the Soviet Union. All of these factors together brought about
the cruel necessity of rationing food staples: bread, fats, sugar, and so on.

Nobody could buy anything without special coupons, which gave the right to buy a certain quantity of goods and food.

We were surviving, if only from hand to mouth, mainly thanks to Torgsin (the All-Union Association for Trade with Foreigners). They would take my mother's few remaining personal possessions—old gold coins and jewelry—and give us in exchange flour and sugar, the necessities of life. But we still lived in want.

At first I was a poor learner. I fidgeted, chattered in class, read unauthorized books, and fought with the others. Some of my friends and I were strangers at this school—that was a principal reason for our constant fighting with the "old" schoolboys, who felt animosity toward newcomers. For talking with classmates and reading literature during class-time I was often sent out of class. I used to wander about the town killing time until I could go back home. The only thing I was good at was recitation, when the teacher asked me to describe to the class the books I had read. I read much more than the other students. Everyone liked listening to me.

On the way to school was the Glebuchev ravine, a conspicuous feature of Saratov. Its scattered hovels were mainly inhabited by Tatars, descendants of the Tatar khanate, the "Golden Horde," who long ago had settled along the Volga. Later, the khanate was divided into the Kazan and Astrakhan kingdoms, both of which were conquered by Czar Ivan the Terrible in the sixteenth century. The Saratov Tatars were the descendants of these "old" Volga Tatars. This district, where my first school was located, had a reputation as a dangerous and unsavory place. We were forever scrapping with the Tatar boys. Our fights sometimes ran to the wielding of "thunder flashes" (homemade pistols loaded with the sulfur from matches).

Much of the time we spent in the street, skating (with the skates strapped to our boots by rope) or skiing. We often went out of the city into the hills or over the ice down the Volga to Engels, the capital of the Volga Germans.[3] In those days we went out in all weather, unafraid of the hardest frosts.

After I graduated from the 4th Form (the American equivalent of late primary and middle school) I went on to secondary school, where I began to make something of myself. It was a better school than the first, and I had grown wiser. After the 7th Form I was even awarded a commendation. I was particularly good at verbal subjects (such as history, geography, and literature). I was not bad at mathematics and physics, though not adept enough at either subject to think of choosing one as my profession.

From the age of six I studied German. Despite our shortage of funds I studied with a German lady for several years, almost until the outbreak of World War II; this was no doubt due to my mother's influence. My mother considered that each intelligent man had to know at least one foreign language. She herself knew French and German. (The latter was very popular in Russia then, before the war with Germany began in 1941.)

Until his arrest I was with my grandfather nearly every day. This had a great effect on me. He read to me a lot in Russian and Ukrainian, especially on Ukrainian history. He told me about ancient Rome, and instilled in me a love of mathematics, setting me a number of problems to do with his work at the Department of Weights and Measures. In late summer and autumn I used to go with him to the wharf to buy melons. We bought them by the sackful. I learned from him how to tell the different steamers apart, even by their sirens (in those days many of them used to run up and down the Volga, and we easily saw them from our enormous veranda).

In 1939 with my mother and father I made my first trip down the Volga to what was then Stalingrad. My father had been thinking of moving there, but somehow nothing came of it.

In summer we usually went away from Saratov: twice or three times we traveled to Odessa where we lived in the dacha—a summer residence or cottage—belonging to my grandmother's brother; once we stayed with my grandfather in Gadyach (Ukraine); and occasionally went to Moscow or the Moscow area. Once or twice we stayed at Istra, a small town on the river of the same name, and in 1940 in Losinoostrovskaya (part of Moscow, where my two daughters now live).

In 1939 the Finnish war broke out. There was a terrible frost. All at once there was a bread shortage, and long lines started to form in which I had to stand every day. It seems strange to look back on it now. Sometimes the bread "ran out," and people stood for hours in the cold waiting for something to arrive—rolls, pies, or *kukh* (from the German *kuchen*, meaning buns or pastries).

During these years before the Second World War, while I was still in high school, I began to attend my father's shorthand classes, at my mother's insistence, and eventually I became qualified as a stenographer. My father felt this profession would be useful for me in case I used shorthand widely, especially at the institute. There was a shortage of textbooks at that time, so being able to write down comprehensive notes in shorthand was a useful skill.

With the outbreak of the Great Patriotic War, which the Soviets called the war with Germany in 1941–1945, our difficulties worsened. Father was not called up on account of his age. From July or August the German air raids began on Saratov, although at that time they only tried to bomb the railway bridge over the Volga, which was of strategic importance. Schoolboys were sent off to dig trenches; they practiced stripping and assembling rifles; they learned bayonet fighting. They were also taught to drive tractors, and in summer and autumn they were packed off to collective farms. At that time all the ethnic Germans were deported, and I used to see abandoned villages along the Volga in which the hearths were still warm. Several times I even gave mock "lectures" before the start of the daily cinema shows on "How to blow up German tanks with Molotov cocktails."[4]

We were starving once again. We had no vegetable garden like some other people, and had nothing to do. My grandmother busied herself by swapping household goods and books for produce in the market. If she was lucky we were able to "feast" for a few days, and we managed sometimes to send parcels to my grandfather, who suffered from tuberculosis, in the prison camp.

Because of the influx of evacuees we used to study in three shifts. In one class I sat with several children of actors from the MKhAT theaters (the Chekhov Art Academic Theater in Moscow), with whose help I was able to see two or three shows (in spite of the war it was very hard to get into the MKhAT). One of them was *The Three Sisters*. A number of brilliant actors from several of the music theatres in Moscow had been evacuated to Saratov; I was therefore able to attend nearly all of the best-known operettas at the opera house. I was also deeply impressed by the officers from the Polish liberation army whom I met at these shows. Their uniforms and their bearing were unusual and not seen in our own commanders (we had "commanders" in those days). It had been drummed into us that all officers were ruffians and drunkards. The Polish army stayed in Saratov for a few months and then departed via Iran for Egypt. They fought against the Germans alongside the British. (They should not be confused with the Polish Contingent, which was formed in 1943 on Soviet soil.) During the time when the Poles were staying in Saratov the local radio used to give broadcasts in Polish.

After all this, the mass shooting of Polish officers by the NKVD outside Katyn (near Smolensk) struck us as particularly savage. The massacre was shortly followed by the announcement that our own army had allowed the Germans to crush the Warsaw uprising in 1944! There had already been some talk of this, as my father and I had been closely following the events in Poland in 1939 which touched off World War II. The German blitzkrieg on Poland in literally two or three weeks was touted as one of the most brilliant operations in military history! They merely forgot to add that the final betrayal, the "stab in the back" to Poland, was delivered by the Red Army, the self-proclaimed "liberator" of western Ukraine and Byelorussia.

As my call-up age was approaching, my mother feared that I would find myself in the army without having completed my secondary education (as if my school certificate would have much importance at the front!). So she enrolled me in a correspondence course in which I simultaneously studied both Form 9, like my schoolmates, and Form 10. Thus I received my certificate in the summer of 1942, a year earlier than my schoolmates.

My father seems always to have wished to leave the past behind and to get away from Saratov, the city of his banishment. It was this, rather than the danger of the German advance already closing in on Stalingrad, that made him take advantage of an opportunity to leave. Through some people he knew he obtained permission for us to depart for Kzyl-Orda—a town in Kazakhstan, a provincial center on the Syr-Darya River, not far from the Aral Sea.

We had to give up everything—our flat, the furniture, many personal possessions, and our books. I don't remember how we got there, but we eventually found ourselves in a "dump" a thousand miles from civilization, although it was the center of the Kzyl-Orda Oblast. We found accommodation with a Kazakh, a watchman at the slaughterhouse.

The evacuees in general were treated quite decently. My father organized his shorthand courses—a skill thought necessary in the Soviet Union even during the war, as not many people knew shorthand in the country—and he and my mother began to earn a little. Food was strictly rationed: during the war a critical shortage of food and other necessities made rationing unavoidable. Russia had lost, during those years, the principal grain-producing regions, the North Caucasus, Ukraine, and Kuban. Besides, we had a huge army to feed. This caused hardship for everyone. Most important for our family in Kzyl-Orda, we obtained bread coupons (the other coupons we had were basically useless since there were no other foodstuffs for which to use them, and we couldn't trade them for more valuable coupons). For an additional kilogram of bread per week my grandmother used to teach the landlady's sister reading and writing. It was my job to go to the market two or three kilometers away for firewood (the wood was from the saxaul tree, which grows in the deserts of Central Asia, and is very heavy).

At that time the Combined University of the Ukraine was set up in Kzyl-Orda, consisting of remnants of the teaching staff from the universities of Kiev and Kharkov. I enrolled for physics and mathematics, though I did not study them for long, and transferred—with a lot of trouble, as they hung on to every student—to the Crimean Institute of Medicine, which had been evacuated from Simferopol. I was happy to make this change: I had developed an interest in medicine through reading many books about medical men, and I had also learned that, for me, mathematics in its own right, or as an academic discipline, was a tiresome branch of science!

People are united by misery and hardship; it helps them to survive. We got along amicably and, as far as possible, happily. I made some friends and my first girlfriends from among the female evacuees. I joyfully received the rare letters that arrived from my schoolfriends in Saratov.

Kzyl-Orda had no buses, let alone trams, and we had to walk every-where. Despite the war it was relatively peaceful and you could even walk about at night.

In December 1943 I turned seventeen. Within a month I received my draft notice. A few days later I turned up in Tashkent at the army school where the signalers were trained. But before being dispatched to the front I was summoned before the garrison medical commission, who graded me as unsuitable for active service because of the residual effects of poliomyelitis. After returning to Kzyl-Orda I resumed my studies at the Medical Institute. I did well, and after my second term I was awarded a Stalin scholarship (which was much higher than the other scholarships).

In the summer of 1943 my father contracted typhoid fever, and just as he was recovering he suddenly died. Since leaving the area I have never returned to his graveside. I have had to travel about a great deal, but I have never gone back to Kzyl-Orda. Now, because of the great distance and my own uncertain health, it is out of the question.

At that time my mother was only forty, but she never remarried. In the early winter of 1944 she went back to Saratov with my grandmother. I followed them after completing my third term. I traveled for several days, standing in the train's unheated corridor (in winter!), hungry since "traveler's" bread coupons—special rations designed for travelers—were not being issued. I was almost penniless. Only one branch line connected Uzbekistan and central Russia, and the trains were always crowded. Even if I had had the money, it would have been very difficult to get a ticket even for sitting on the hard seats in the carriages. I had to change trains at Iletskaya Zashchita and wait a whole day for the next train.

There I ran into difficulty. During the war there were outbreaks of spotted fever, or typhus, which is transmitted by lice, and all travelers had to submit to—and pay for—louse checks at the railway stations. I had enough money for either the check or for a railway ticket, but not both. I had to spend my last money on a medical check because I couldn't travel any further without one. But I had no money left for a ticket.

As it happened, I was near to the head of the line for ticket punching. A passenger traveling on duty did not want to wait in line and asked me to get his ticket punched for him. He gave me some money, which was enough to pay for my ticket as well. I found myself back in the freezing cold corridor!

In Saratov I registered for the second course at the Medical Institute. Since there had been no official evacuation of Saratov and we had left of our own accord, we naturally failed to regain our own flat. Luckily, a relative of my mother took us in. I owe her many thanks. She herself lived in meager circumstances. After the arrest of her husband and son-in-law she stayed in a two-room communal flat, together with her daughter and two small grandchildren, and now there were three more of us to house! After about two years of living under these conditions my mother somehow managed to get a small room in the same building. From then on it was like living in paradise.

I was doing well at the institute. My friend Sasha Greten (who is now a professor at the Medical Institute in Nizhnij Novgorod) and I always passed our exams ahead of schedule. In my third year I branched out into pathological physiology with the strong support of Prof. O. S. Glozman, who had instilled in me a taste for scientific work. Later on Sasha and I switched over to therapy, because I had always dreamed of becoming a physician. The holder of the chair of diagnostic medicine, Dr. Simagina, presented me with a book bearing the inscription, "To Igor Domaradskij, a physician by vocation."

In Saratov I had to join the Young Communists—a necessity, or I would not have had any prospects of doing scientific research. After that I was accepted as a candidate for membership in the Communist Party and by 1946 I had become a Party member. Naturally nobody knew anything about my background or the legal proceedings against my relatives. I somehow managed to keep it quiet.

I can't say exactly what made me join the Party. Two main factors were responsible for it. First, there was some curiosity as to why I hadn't become a Young Communist, which might have raised some unpleasant questions since in those days everyone on reaching the age of fourteen almost automatically joined the Komsomol.[5] For me, joining the Komsomol had never been an option. While I was at school in Saratov, before we left for Kzyl-Orda, I was not accepted for the Komsomol because of the charges of anti-Soviet activity leveled against members of my family. Second, having once agreed to be part of the Young Communists, I had to become a member of the party or risk suspicion. At my age I would have been expected to apply for candidacy to the party, especially since I was considered a good student and hence conspicuous. Despite my distrust and fear of the party I was well aware that my future largely depended on party membership. I was fed up with living like a semipauper. Also, I was hoping to shed my past; despite the death of my mother's father, both my mother and grandmother supported me in this. At this point in my life, I cared less for being a troublemaker, a secret rebel, than for establishing myself in some profession. I consoled myself with the thought that the longer I was in the party among respectable people, the less chance there was of bringing up the past.

I realize now that this was a naive attempt at justifying my membership!

NOTES

1. An artel was a type of teamwork organized on a voluntary basis for some form of communal activity, such as for fishing, cutting firewood, and hunting.

2. Unlike the practices found in most Western nations, the Soviet Union required its citizens to possess passports not only for travel outside that nation but also for internal travel. In the case of the collective farmers, this State-sanctioned practice of denying them passports had the practical effect of forcing them to remain on the farms and to continue to be employed as farm workers. (Ben Garrett)

3. German emigrants resided on the Volga (between Samara and Saratov) in the eighteenth century, from the time of Catherine the Great. Under the Soviet regime a German autonomous republic was established. The capital was Engels town. The Volga Germans were expelled to Siberia and Kazakhstan at the beginning of war with Germany in 1941.

4. A "Molotov cocktail" is any bottle containing flammable liquid, such as alcohol or gasoline, and having a rag or other material hanging out the mouth. The rag is set on fire, the bottle is tossed, and a small fire is produced when the

bottle breaks on impact. This low-technology approach to warfare was pursued by the Soviets during World War II and proved effective. The technology and the name remain with us today. The name of the device honors Vyacheslav Mikhailovich Molotov (1890–1986), who served as the Soviet Foreign Minister during the war. Born "Skryabin," he took the name "Molotov" (Russian for "mallet") in 1906 as his pseudonym or Party name. He was a relative of the famed Russian composer Alexandr Nikolaevich Skryabin (1872–1915). Molotov fell from favor with the Communist Party following Stalin's death in 1953 and was expelled from the Party in 1964. He served out his career thereafter in relatively obscure or insignificant posts (Ben Garrett), including that of ambassador to Outer Mongolia (see chapter 4, "In Siberia."—W.O.).

5. The Young Communist organization.

CHAPTER 3

"FIRST STEPS ALONG THE WAY"

Neither faith, nor law.
(from the Russian)

I graduated from the institute in 1947 with honors and I could easily count on an internship at a medical clinic. But at that time graduates who had served at the front received preference. Eventually, I was appointed to one of the country hospitals in the Saratov Oblast, and it was only by a miracle that I obtained a research post in the postgraduate course of biochemistry, which I have never regretted. This was good luck; it was also my one opportunity to devote myself to science.

In 1946, to my mother's alarm, I got married. My wife, Natalya, was eighteen months older than I. She had graduated earlier from the institute as a pediatric physician and had gone off to her parents in the small town of Kirsanov in Tambov Oblast because there was nowhere for her to live in Saratov. There she found a job in the medical section in the village of Bibikovka in the Penza Oblast.[1] Her father wanted to hear nothing of my research post. He insisted that in order to support a family I would need to earn a living as a medical practitioner. Incidentally, many years later I happened to learn that my then father-in-law, like my father and grandfather, had also been the target of an NKVD proceeding, and that was why he found himself exiled in Kirsanov.

Shortly after I graduated from the institute my wife moved back to Saratov, where at first we lived with my mother and my grandmother in a very small flat. This marked the start of our struggle to secure our own flat. As a doctor my wife was earning pennies, and I even less.

Even today, the salaries of medical doctors, teachers, and science researchers in Russia are extraordinarily low by Western standards: $30 to $60 per month. It is essentially equal to minimum wage or a little more by Russian standards. That is a main cause of discontent and strikes. So it was always. In the Soviet Union manual labor had priority as compared to mental work. This can be understood if we remember that the "hegemony of the proletariat" was part of Soviet Marxist ideology. But now this lack of respect and the terrible wages are some of the principal causes of the "brain drain." Many science researchers live as emigrants today in the United States and in Europe; among them are a whole series of my students and coworkers. Some other science researchers have been forced to leave their work and look for a new occupation in Russia (e.g., one of my daughters and a son-in-law) because they cannot make a living as researchers. But in Soviet days such emigration was impossible.

Our life was further darkened by my mother's illness; she had been diagnosed with a serious form of pulmonary tuberculosis, from which both my aunt and grandfather had died. Luckily there was now a new medicine called aminacyl, and the antibiotic streptomycin was also available. Thanks to the kind intervention of V. V. Akimovich, a faculty member in the Department of Microbiology, his wife, a pulmonary specialist, had my mother back on her feet relatively soon.

Life became somewhat easier only after the abolition of the coupons system and the exchange of old money for new, devalued currency in 1948. But my hunt for a private flat for my wife and me went on for several years, until finally the mother-in-law of my friend V. I. Rubin sold us a small room not far from the Volga. I have no idea how she managed it, because the selling of flats was prohibited then in the USSR. All living quarters in cities belonged to the state, and their sale was therefore prohibited under the Soviet constitution. Village habitations, as well as dachas (country homes) were exempt, but their owners could sell them only with permission of the local governing institutions, and to obtain such permission was virtually impossible.[2]

In the summer of 1950 I finished my research appointment and was slated to be sent to some secret organization in the Urals. (I learned later that this outfit was working in the field of atomic energy, which was then going at breakneck speed.) All such assignments depended on the Third Main Directorate of the Ministry of Health.

This appointment required the appropriate formalities, or, in simpler terms, the clearance for secret work. Without such clearance, I would be idle without a job or income. In my disappointment I wrote to the Ministry of Health asking to be sent to the All-Union Antiplague Institute known as *Mikrob* in Saratov, where the subject matter was related to that of my candidate's thesis. To my surprise, this request was quickly granted. I was accepted for the institute at the grade of junior scientific worker, and in December of 1950 I defended my dissertation on the metabolism of amino acids in *Proteus vulgaris* (the name of one very widespread bacterium), after which I became a junior worker in science, a post rather like that of an advanced graduate student in the United States. I hoped, on the strength of my research, to have the possibility of investigating dangerous bacteria, including *Yersinia pestis,* the agent that causes plague. This appointment redirected the entire course of my career.

As things turned out, the Mikrob Institute at Saratov was only the first step on a long road: I devoted the next twenty-three years of my life to the plague-control system. The Bolsheviks had established the Mikrob Institute in Saratov soon after the October Revolution in an effort to control the continual outbreaks of this deadly infection, in particular in the southeastern part of the country.

Plague, the Black Death of the Middle Ages, is a disease long feared in Russia. During the reign of Empress Catherine the Great, a huge outbreak of plague killing hundreds of thousands threatened Moscow (1771). More recently, epidemics of plague arose in Odessa on the Black Sea and neighboring areas (recorded occurrences include 1812–1813, 1823, 1829, 1835, 1837, 1901–1902, and 1910). The most extensive were the first epidemics when 2,655 people died out of a population of about 25,000 townspeople. It is important to emphasize that all these epidemics in Odessa arose from plague infection brought by ships from Egypt, Turkey, and Greece. In the period 1877 to1879 there was an outbreak of plague in Vetvlyanka on the Volga River, in the Astrakhan region. The origin of this outbreak remains unknown.

The most serious epidemics, however, were of pneumonic plague, which killed 60,000 people in Manchuria (near the Russian frontier) in 1910–1911 and several hundred in the Russian Far East (in Vladivostok and Ussurysk) in 1920–1921. Pneumonic plague is caused by the same bacterium (*Yersinia pestis*), but unlike the bubonic form, which is spread by fleas, pneumonic plague is a respiratory infection that spreads directly from person to person through respiratory droplets.

Though there has not been a major outbreak in Soviet territory in a long time (according to inaccurate official records from 1920 till 1989 when 3,639 people fell sick with plague, 2,660 of these cases were fatal), the threat of plague remains a serious concern. More than eighty thousand acres of the USSR had large populations of rodents—marmots, susliks (ground squirrels), gerbils, and others—where plague infection perennially simmered and sometimes flared into epizootic* outbreaks. These areas are known as natural plague foci-locations where plague circulates in animal populations, usually among animals that have some resistance to the infected fleas; sometimes the germs move from resistant to sensitive animal species, and major epizootics develop. When these epizootic outbreaks sometimes spill over into human populations, they need to be watched. All human epidemics have their start from such rodent epizootics. The causal agent, *Yersinia pestis,* is transmitted from sick rodents to human beings through fleas. Epidemics caused by marmots or tarabagans are thought by Russian plague experts to be especially dangerous: contact with sick marmots or their fleas seems to have started the first Manchurian pneumonic outbreak in 1910–1911.

The initiative for founding the Mikrob Institute came from academician D. K. Zabolotnij, but the idea was implemented with the active cooperation of Prof. A. A. Bogomolets and the energetic Prof. A. I. Berdnikov,

---

*A disease outbreak among animals comparable to human epidemics.

who at that time (1919) held the chair of microbiology at the University of Saratov and who then became the first director of the institute. In 1920 the institute became independent of the university. By 1922 all of the plague-control laboratories in southeastern Russia were combined under its aegis. The institute organized investigations of the steppes and the deserts, and discovered the epizootics of plague among the rodent populations. Eventually the system shifted its emphasis from controlling plague outbreaks to preventing them in the first place. After the Third Pandemic, the greatest outbreak of plague since the Black Death of the Middle Ages, preventing plague outbreaks in the Soviet Union was a matter of singular importance. By 1934, probably because of its superior faculty and distinguished founders, the Mikrob Institute had become the nerve center of the entire Soviet plague-control system, which included facilities in Irkutsk, Rostov-on-Don, Alma-Ata in present-day Kazakhstan, and later Stavropol (in 1952). Plague control in the old Soviet Union grew into a specialized preventive service, the first such organization in the world. While initially Russian specialists such as Zabolotnij had contacts with other plague experts around the world, including the great Chinese plague fighter Wu Lien-teh, and the Austrian doctor/field researcher Robert Pollitzer, over time Soviet experts grew more and more isolated, as contact with outside researchers was discouraged or actively forbidden. The Stalinist government viewed foreign biologists as corrupting and dangerous outside influences.

Soviet plague-control researchers initially carried out their work under complicated field conditions: they were required to serve as microbiologists, epidemiologists, zoologists, parasitologists, and sometimes even general practitioners. These workers were heroes who daily risked their health and their lives: at that time the only way to treat plague was with a partially effective antiplague serum similar to that developed by the Russian-Jewish bacteriologist V. Haffkin.[3] This serum had limited effectiveness, and even then only against bubonic plague, because it provided some immunity against blood-borne infections introduced by fleabites, in cases where the microbe had not widely spread in the victim's body. Immunity against plague is not humoral—that is, produced by antibodies—but cellular. Antibodies carried in serum, therefore, do not directly affect plague bacteria, though they may help the body's own phagocytes in that role. The Haffkin serum, furthermore, did not produce the mucous membrane immunity on the surface of the lungs needed to provide immunity against pneumonic (respiratory) plague infection. Against the more deadly pneumonic form, therefore, it was useless. Not until the late 1940s was real progress made in the treatment of pneumonic plague. In the Soviet Union, Nikolay N. Zhukov-Verezhnikov, together with members of the Mikrob Institute, proposed what was called the "complex method," based on parenteral injection of an alkaline solution of sulphidine with methylene blue, a dye whose precise function remains somewhat obscure, along with the antiplague serum. Streptomycin, discovered by the American scientist Selman A. Waksman in 1944,[4] became available later on; this dramatically improved the outcome for pneumonic plague, which in prior centuries had been almost uniformly fatal.

The history of the fight against plague contains many pages of heroism and superb performance, which will be forever associated with Russia's doctors, medical students, and their support personnel. Their names are inscribed on the roll of honor of the battle against plague. But the story of their heroism is largely unknown, even in the wider medical community. One of these remarkable men was Ippolit A. Deminskij (1864–1912). He was infected in the village of Rahknik (not far from Astrakhan on the Caspian Sea) with plague from a suslik. Sensing the approach of death, he dictated a wire asking for his body to be opened in order to isolate the culture of the plague microbe. He thus proved by his own death that suslik plague was identical to human plague.

Another remarkable physician, Vasily P. Smirnov (1901–1976), was inspired with the idea of vaccinating humans against plague through conjunctiva.* While working in Mongolia, where he vaccinated more than 100,000 people, Smirnov infected himself with plague in order to demon-

---

*The conjunctiva, or membranes covering the whites of the eyes, are connected to the nasal cavity. Liquid vaccine placed in the eyes travels to the nose and into the lungs. But conjunctival vaccination may be dangerous to the eyes. Therefore, instead of conjunctival vaccination, direct nasal vaccinations are now used in Russia. Even so, V. Smirnov's experiments with conjunctival vaccination had great scientific significance.

strate the effectiveness of his method. Control animals died, but he recovered, and thereafter worked for a long time in Irkutsk, where I had invited him, after he was ousted from the plague-control institute in Stavropol. He apparently had some conflict with the director of the institute.

Among the plague specialists, perhaps our last victim of plague was the outstanding scientist Abram Lvovich Berlin (b. 1903), who had worked for some years in Mongolia and had even been to Tibet, the result of which was his remarkable article on "Tibetan Medicine and Plague." During the last years of his life he was deputy director of the Mikrob Institute. In 1939 A. L. Berlin was infected with plague in his laboratory. Unaware of his illness, he traveled to Moscow, where he died, after infecting his barber and some of the medical staff that had been in contact with him. The tragic death of A. L. Berlin brought about new and strict rules for working with plague and other highly dangerous infections. These rules have proved to be fully justified and are largely still in force.

But danger stalked the plague specialist not only in the field or the laboratory. Many of them, including Professor Skorodumov, were also persecuted, and some perished in the cellars of the government secret police. Any outbreak of plague might serve as a pretext for arrest on charges of deliberately spreading plague, or of recklessness. This was a function of the tactics of universal terror so characteristic of the Stalin era: terrifying and persecuting some scientists (or members of any other profession) served to keep the rest in line. It is perhaps most difficult for Westerners to understand that many of these scientists had done nothing politically compromising or suspect even in Stalinist terms. They served only as examples to terrorize others.

The suffering of these plague fighters always haunted me, though for many years I could only remain silent about it. But some years ago, during the era of Gorbachev's perestroika (beginning in 1985), I began to collect information on members of the plague-control system who had suffered during the years of Stalin's persecution. I hoped to rehabilitate those who were still under a cloud. I did this not only in memory of the persecuted scientists themselves, but also in memory of all the victims of Stalin's terror, including my own relatives. For me this rehabilitation effort was a form of protest against the Soviet regime. It has proved extremely difficult.

First, none of the institutes hold any precise data about the number of subjects of Stalinist repression, possibly because the relevant documents have already been seized by the security service. Sometimes it was the families of those who were persecuted who destroyed the documents. Other family members may have sought to protect themselves from guilt by association. Second, many of these former victims were not working in the institutes but on plague-control stations or in departments widely scattered over the vast tracts of the country. Many of these stations are now closed, while at others there is no one who can recall the events of that distant past. Third, many witnesses of the events of these times have not yet recovered from fears for their own safety. They have no confidence in the permanence of the current political change and the final disappearance of the Soviet regime. Either they pretend to know nothing or they frankly avoid any discussion of the subject. Even directors of some institutes—the "young" who either didn't care about such things or weren't convinced of the innocence of the victims—avoided answering my questions about the fate of their predecessors. In any event, my efforts were obstructed. But I managed to gather some evidence to help clear the names of colleagues, for which I am truly grateful to my respondents. I am obliged to the former director of the Mikrob Institute, Prof. A. V. Naumov; Prof. L. N. Klassovskij; Prof. I. F. Zhovtyj (recently deceased), my deputy in Irkutsk; and to Dr. L. A. Avanyan (in Stavropol).

May the bright memory of all those who innocently suffered, whoever they may have been, whether professors or workers, be preserved forever in the hearts of those who live and who will come after us.

Two years after defending my dissertation I was made a senior scientific worker at the Mikrob Institute, and soon received the title of assistant professor or a senior research officer, certified by the State High Certification Commission (VAK in Russian). I immediately could command a higher wage and, borrowing the money I lacked, I bought myself a motorcycle (one of the cheapest). It brought me no luck. I soon had a collision with a truck in the dark and thereafter I began to "limp in both legs." Later I was to enter this on questionnaires under the heading of "Distinguishing Marks."

Since the biochemistry of the plague microbe had not yet been studied at all, I chose to study the protein metabolism of this organism as the subject for my doctorate. I worked hard and by 1955 I had already collected the material I needed. In addition, I worked with tularemia and brucellosis.[5] I caught brucellosis and had to spend a couple of months in the hospital. Just at that time my grandmother died at the age of seventy-two. I was too ill to attend the funeral.

A few years earlier, I had also contracted malaria. Both before and after World War II many people came down with it. Perhaps this is why every summer my parents used to take me out of Saratov and well away from the mosquito-ridden Volga River, whose beauties I came to appreciate only much later and which I then never even suspected.

Despite the improvement in my own circumstances, a new wave of repression had begun in the late 1940s after the "Leningrad Affair" (1946–47). (This was the crackdown on the Leningrad party organization caused by the infighting among top party echelons, in which almost all top party leaders and officials in Leningrad were executed.) The "Leningrad Affair" was a continuation of an intra-party fight of which I do not know the details, only the result—more terror. At the same time a new wave of persecutions began, this time of intellectuals. In my circle, my friend Yuri Demidov's sister was arrested. You will recall from chapter 1 that his father, the barber, had been arrested in the 1930s. At that time Yuri was working as a magistrate near Saratov, where he was sacked from the police. For some years afterward, and up to the "Khrushchev thaw" (a short period after Nikita Khruschev came to power in 1953) he had to travel through the district as a lecturer for the Society for the Propagation of Knowledge. But despite all his misfortunes he kept his spirits up and subsequently became a distinguished criminologist, a professor and, eventually, a colonel of the police.

Some of the repression was aimed at scientists, specifically at those who did not follow entirely the party line on "Soviet science." Geneticists and other biologists were particularly at risk, because of Stalin's peculiar devotion to the doctrines of Trofim Lysenko (1898–1976). It was Stalin who played the decisive role in the rise of "Lysenkovshchina" or "Lysenkoism."[6] Stalin divided science into two categories: his own brand, "materialistic" and "progressive," and everything else, "bourgeois, outmoded pseudoscience." Lysenko became Stalin's active assistant. Promising prosperity for our collapsing agriculture, Lysenko created a new form of genetics based on the "research" of biologist-selectionist Ivan V. Michurin. According to Lysenko, two dogmatic assertions underlay this "theory." First, he asserted that changes to the body during the life of an individual organism could be transmitted to its offspring. In other words, the long neck of the giraffe was acquired by generations of giraffes stretching their necks out to reach taller and taller trees. This is a theory known in the Western world as Lamarckism, after Jan Lamarck (Darwin's predecessor). Second, Lysenko discounted the work of the Austrian monk Gregor Mendel, who demonstrated the existence of hereditary particles known as "genes"; Lysenko in general denied the very existence of genes.[7] Thus he rejected both Mendelian genetics and Darwinian evolution.

Rejection of the fundamental principles of modern biology could not fail to have a disastrous impact on Soviet science, and so it was. The session of the All-Soviet Union Agricultural Academy (VASKhNIL), named after the Bolshevik revolutionary leader V. I. Lenin, in August 1948 was a triumph for Lysenko, because the text of his speech at the session was previewed, edited, and approved by Stalin personally. After that, all opponents of Michurin's biology were accused of being idealists and metaphysicians, worshiping foreign ideas. Their studies and research were considered to be in conflict with the basic platform of the party and the government. As a result, thousands of biologists, researchers, and teachers were expelled from their positions and replaced by ignorant or unprincipled people. Texts and scientific works containing materials contradicting Michurin's biology were withdrawn from the libraries, and sometimes destroyed. At the same time the session of VASKhNIL opened the way for the "new cellular theory" of Olga B. Lepeshinskaya and such scoundrels of science as G. M. Boshyan, G. P. Kalina, and others. A perverted form of Ivan Pavlov's physiology also was associated with Michurin's biology.

Thus, "Lysenkovshchina," as the Michurin-Lysenko doctrine is known in the former Soviet Union, and all that flowed from it, harmed the mentality of Soviet scientists beyond description. Threatened by violence or tempted by benefits, the scientists were induced to act amorally in the extreme.

An incident with Aleksandra A. Prokofyeva-Belgovskaya, a famous geneticist and opponent of Lysenko, represents a striking example of how Lysenko and his cronies interfered with scientific research. Prokofyeva-Belgovskaya prepared a doctoral thesis, but it was at variance with the principles of "Michurinian biology." She fell immediately afoul of the system. In the USSR, after a public discussion of a Doctor of Science Thesis at the Academic Council of an Institute, the results of the vote and all the required documents were examined by the governmental Higher Certification Committee, which was run, as a matter of course, by the Communist Party. Only after that vetting was the candidate awarded the Doctor of Science degree. The chairman of the Higher Certification Committee told Prokofyeva-Belgovskaya that the content of her thesis did not agree with the principles of "Michurinian biology." However, since the members of the committee knew her as a talented scientist, and since "Lysenko himself" had been following her work, the committee proposed that she add only one phrase to the summary of the thesis. The chairman suggested she add that all the experience of her work in the field of genetics had led her to the conclusion that "Mendelism-Morganism" was erroneous and that "Michurinian biology," as developed by Lysenko, was correct. Prokofyeva-Belgovskaya was given the opportunity to think about this and come to the next meeting. She immediately and firmly rejected this offer and said that, on the contrary, all of her scientific experience had allowed her to prove the correctness of the chromosome theory of heredity developed by Thomas Hunt Morgan[8] and his disciples. After that the USSR Higher Certification Committee voted down the thesis of Prokofyeva-Belgovskaya. She was fired from the Institute of Genetics in 1949. Only in 1965 did A. A. Prokofyeva-Belgovskaya receive her Doctor of Sciences degree. This happened when Khruschev lost his position as leader of the Communist Party and of the government, and Lysenko, losing thereby his personal protection, was relieved of his position as director of the Institute of Genetics and of his duties as a member of the USSR Higher Certification Committee.

In 1948 I attended an open party assembly at the Saratov Medical Institute devoted to the famous session of VASKhNIL (July 31–August 7, 1948). The principal "scapegoat" there was one Professor Lunts, who was declared to be a "blatant advocate of Mendelism and Morganism" and an adherent of the doctrines of the noted German philosopher Georg W. F. Hegel, which had been rejected in the USSR as reactionary. He had admitted the charge himself in reply to a question from the chairman of the meeting, Professor Papovyan, who had been specially sent to Saratov by the Central Committee of the Communist Party of the Soviet Union and the Ministry of Health to be the principal of the Medical Institute. Papovyan, "to impose some order," demanded, "Who are you, Professor Lunts?" Papovyan's question had a specific political meaning. It was intended to force Professor Lunts to plead guilty to the charge of opposing Lysenko and adhering to the reactionary Western doctrines of Morgan and Mendel.

Both of my dissertations were necessarily written, therefore, under the requirements of various government decrees, which it would have been impossible to disregard[9] when writing a dissertation. Every research paper had to contain references to party and government decrees on ideology, regardless of their subject matter (Ivan Pavlov's theory, Trofim Lysenko's work, or even Stalin's work on linguistics). My candidate's thesis,* for example, had to reflect the fight against cosmopolitanism, or adulation of the West. The party and the government tried to isolate us, in the belief that as scientists we should be completely self-reliant. An examiner insisted that I insert a reference to "nervism" in my work on the biochemistry of one of the bacteria! I also was obliged to add references to the "doctrines of such standard bearers of science" as the old Bolshevik Olga Lepeshinskaya, her successor G. P. Kalina, and the scientific char-

---

*There are two scientific degrees awarded in Russia. The first is the candidate degree (for example, a candidate of medicine or chemical sciences). The second (the highest) is a doctoral degree (for example, I am a doctor of medical and a doctor of biological sciences).

latan G. M. Boshyan who "discovered" the ability of bacteria and even antibodies to turn into crystals and back again.

But "Lysenkovshchina" was far and away the most pernicious doctrine. If any facts failed to fit in with this "progressive" evolutionary doctrine, they were either to be treated "appropriately" or rejected. As it developed, influential members of the All-Union Antiplague Mikrob Institute—Galina Lenskaya, my own chief; Prof. N. N. Ivanovskij; and the quite well known N. N. Zhukov-Verezhnikov, who had together developed the "theory of the conversion of the plague agent into pseudotuberculosis"—were blatant adherents of Lysenko. The theory of the conversion of the plague agent into pseudotuberculosis is contrary to Darwin's doctrine—any species arises in the course of evolution only once and not every day, and not suddenly. The so-called theory of my colleagues was based simply upon speculation and guesswork.

On the one hand there were many who realized the absurdity of all this nonsense, while on the other hand all of us were forced to condemn and unmask all "hostile attacks" on these "progressive theories." A pharisaical atmosphere prevailed, from which, as a Party member, I found it difficult to keep my distance.

While working at the All-Union Antiplague Mikrob Institute I also came up against "Jewish" history, which might quite well have affected me: my grandmother was Jewish and many people knew her. This history is of some interest as it also involved the "case of the poisoning doctors" (known in English as the "Doctors' Plot").

Anti-Semitism has been manifested in many ways for as long as Russia has existed. It has been most prominent since late in the eighteenth century with the incorporation into the Russian Empire of western territories with a large Jewish population. From about that time most of Russia's Jews were restricted to living in a circumscribed area termed the "Pale of Settlement" (*cherta osedlosti evreev*). Many Jews managed to leave the Pale and migrate to other parts of the empire, but they were subjected to periodic expulsions from forbidden areas, and it was not until the fall of the tsarist government in 1917 that this form of apartheid was fully eliminated. Meantime, Jews were subjected to quotas in admission to various public institutions; they were also barred from the civil service and the officer corps.[10] Not by accident, therefore, we find that among the leaders of the Bolsheviks there were so many Jews (Lev Trotsky [Bronstein], Yakov Sverdlov, Grigiriy Zinovev [Rodomyslensky], Lev Kamenev [Rosenfeld], and others). But it is worth noting that all of them were killed later, during intra-Party struggles.

During the Soviet period, everyday anti-Semitism remained alive and well in Russia. Beyond that, the horrendous pogroms of 1918–1921 took a minimum of 50,000 lives (although most of the perpetrators in this case were probably Ukrainian, not Russian). From November 1948 onward, the Soviet authorities conducted a deliberate campaign to liquidate what was left of Jewish culture. The Jewish Anti-Fascist Committee was dissolved and its members arrested. Jewish literature was removed from bookshops and libraries, and the last two Jewish schools were closed. Jewish theaters, choirs, and drama groups, amateur as well as professional, were dissolved, and hundreds of Jewish authors, artists, actors, and journalists arrested. During the same period, Jews were systematically dismissed from leading positions in many sectors of society, from the administration, the army, the press, the universities, and the legal system. Twenty-five of the leading Jewish writers arrested in 1948 were secretly executed in Lubyanka Prison in August of 1952.

The anti-Jewish campaign culminated in the arrest, announced on January 13, 1953, of a group of "Saboteurs-Doctors" accused of being paid agents of Jewish-Zionist organizations and of planning to poison Soviet leaders. Fears spread in the Jewish community that these arrests and the show trials that were bound to follow would serve as a pretext for the deportation of Jews to Siberia. But on March 5, 1953, Stalin unexpectedly died. The "Doctors' Plot" was exposed as a fraud, the accused released, and deportation plans, already discussed in the Politburo, finally dropped.

The "Doctors' Plot" incident took place during the regime of the Minister of Health, Col. Gen. E. I. Smirnov (from 1947 to 1953), who could hardly have doubted the innocence of distinguished scientists such as Academician Lina Shtern, members of the Academy of Medical Science V. N. Vinogradov, V. Kh. Vasilenko, M. S. Vovsi, and several others who joined the troop of "poisoners." I later got to know Smirnov, although I never had the chance to question him either about this or about whether he had been able to do anything for the rehabilitation of those accused. It is also remarkable that, although all of those who had suffered in the case of the Jewish Anti-Fascist Committee and the "Doctors' Plot" were rehabilitated, either while alive or posthumously, "Point 5" in employment questionnaires remained an obstacle for Jews. No person would be hired without filling out a personal history, which included Item 5 (ethnic origin). Any one whose Item 5 ("Point 5") was "defective" (i.e., that person was a Jew, or someone devoid of most civil rights) could be sure of poor career prospects. "Point 5" cast a long shadow behind it, and Jews by no means took a back seat among the ranks of dissidents. The purity of "Point 5" was very carefully fostered in the Soviet science system, which I will describe below. In the major scientific centers such as Obolensk or Koltsovo there were no "covert Jews" so far as I am aware, so that in private we used to refer to them as "cities without Jews."

The echo of the "Doctors' Plot" reverberated as far as Saratov. Among those I knew who suffered was the Jewish Prof. L. S. Shvarts, a brilliant clinician and therapeutist who, unlike many others, was a genuine scientist. The "plot" served as a pretext for hounding him. It is interesting to note that his two non-Jewish coauthors, Professors N. N. Ivanovskij and A. M. Antonov, were declared "not to blame."

A second Jewish victim was Evgenia Bakhrakh, a member of the All-Union Antiplague Mikrob Institute, who had been an active Young Communist and local Party official. She was later restored to the institute as a Candidate of Science, but only in the post of laboratory assistant, and she was unable to defend her doctoral thesis for many years. In 1952 I happened to be in competition with E. Bakhrakh for the post of senior scientist, which I won, although, under different circumstances, by seniority and work experience the position would certainly have gone to her.

Finally, the third victim was Ya. L. Bakhrakh (a namesake of E. E. Bakhrakh), who was accused of conducting "human experimentation." To obtain the amino acid cystine he had used sweepings from hairdressers for his raw material. This shows the lengths the persecutors of Jewish scientists were willing to go to come up with some sort of basis for their accusations. The absurdity of the charge of "human experimentation" on dead, discarded hair is self-evident. A feature about this subject nevertheless appeared in the satiric magazine *Krokodil*, which was followed by Bakhrakh's dismissal and the rejection of his candidate's thesis by VAK. For a long time the poor man eked out a miserable existence until finally he turned up in Irkutsk. There, I helped him to write a fresh thesis on the part played by cystine in the treatment of bone fractures, which was not defended until 1964.

A wave of anti-Semitism rolled across the entire plague-control establishment; in particular it enveloped the Antiplague Institute in Rostov, where I was later to work. There, the principal victim, among others, was Prof. I. S. Tinker, the deputy director, who had done a great deal to eliminate plague from the northwest Caspian shores and had been directly involved with E. Ya. Elbert and N. A. Gaiskij in developing a live tularemia vaccine. An excellent sketch of I. S. Tinker has recently been written by his son, A. I. Tinker, who is also a professor. He describes how his father was shunned because of his Jewish heritage by many of his colleagues and how, while looking for work, he wrote for an employment interview with Zhukov-Verezhnikov, then deputy minister of health, who knew Tinker well from the plague-control establishment. But he turned Tinker down.

My doctoral work aroused no special enthusiasm in the Mikrob Institute, as very tough standards were then being stipulated for such dissertations. A number of very solid and erudite plague specialists were unable to write one because of the new standards, while I was considered to be too young and impudent to have produced one myself. (The envy of my older colleagues perhaps played a role in this.) In this context I must mention academician A. E. Braunshtejn. What it was in me that appealed to him I don't know, but when I was working in his Moscow laboratory (learning the technique of paper chromatography) he listened to the results of my work at his seminar and gave his approval. After that I had no special trouble in defending my thesis, although, owing to my boss in the All-Union Antiplague Mikrob Institute, Professor Ivanovskiy, with whom my relations had been strained for a long time, the defense was held up for a whole year until 1956. Less than a year later I was confirmed by the VAK and became a doctor of medical science. (One of my examiners obliged me to insert in my thesis one and a half pages of miscellaneous rubbish dealing with the Lysenko doctrine.) After being confirmed I was appointed as head of the biochemical and biophysical department of the Mikrob Institute.

The defense of a doctoral (as opposed to a candidate's) thesis by young people was an infrequent event at that time, and work on it required many years, although now the process has become simpler. The main obstacle on the way to the thesis defense was the necessity to publish articles. Publication was a very difficult task, as there were very few Russian journals, and any publication abroad was prohibited. In my case, furthermore, publications on my subject area, plague, as well as on other especially dangerous infections like cholera and anthrax, were as a rule prohibited. So I was not expected to publish, which helped me to prepare my doctoral thesis relatively quickly (within four years).

It may seem surprising to scientists in the West that so young a scholar—directly after the acceptance of his doctoral dissertation—was appointed to so high a post. But I may have had little competition. The biochemical department was, in fact, organized by my teacher, Prof. N. Ivanovskij, in the late thirties, but he left the institute in the early fifties. The post of head of the department was vacant for two or three years. Most scientific researchers in the Mikrob Institute (and in other microbiological institutes) did not take an interest in biochemistry and especially in biophysics, as neither of these had direct applied significance at that time. Their interests, instead, were in the sphere of microbiology, epidemiology, and zoology. The biochemistry of microbes was almost unknown at that time, at least in the Soviet Union.

Meanwhile my personal life was in turmoil. Endless quarrels had finally led to my parting company with my wife who, without telling me, had taken our one-year-old daughter off to her parents in Kirsanov. I will not claim that right was on my side. Too much water has flowed under the bridge and there is no point in dredging up the past. I will merely say that at that time I had developed a strong, and as I thought, sincere attraction toward another woman which lasted for several years. The romance was broken off on my initiative, and to this day a feeling of guilt toward this lady remains with me.

The break-up with my wife did not ultimately affect my career, though at first I had some serious difficulties. Divorce was considered a sign of moral turpitude and thus was discouraged among members of the Party. Many times I was brought before the district committee of the Party. These sessions ended in fists pounding on the table and threats of "fearful penalities" if I did not return to my family. But the defense of my doctoral thesis, which in many ways was sensational for the time, changed my whole life. Not only did it make my divorce more palatable but also I received a considerable raise in wages.

Once, when I happened to be in Moscow (where I had been sent to procure reagents for our work),[11] V. Zhdanov, the deputy minister of health, who was himself a relatively young man, suggested that I should become the director of the Antiplague Institute at Irkutsk for Siberia and the entire Russian Far East. This suggestion came as a complete surprise, as I had never thought about an administrative career. To this day I cannot explain why the ministry was interested in me. Perhaps it was because I was a doctor of medical science and a young man. Such a combination was rare in Irkutsk, and indeed in all of Siberia. After a long hesitation I agreed to take the post. The ministry promised me freedom of action and the possibility of engaging in scientific work. Of course, it was a risky venture on my part. But I took the risk and never regretted my decision. After consulting my mother on the telephone I accepted the post.

On my return to Saratov I reported all this to D. G. Savostin, the director of the institute, and to Galina Lenskaya, his deputy. Knowing me fairly well, they said I was completely off my head. Lenskaya added that this was a fair way to end up in prison, or in the best case in ruin, "with all the unpleasant consequences." My colleagues believed that holding down such a post would be difficult for me, as I had no experience in administrative work and I would be asked to head an entire institute. I began to harbor doubts as to the advantages of this direction in my life, about which I wrote to B. N. Pastukhov, at that time the head of the section for highly dangerous diseases at the Ministry of Health. My letter opened with the words, "I am very clearly aware that I acted rashly in hastily agreeing to being transferred to Irkutsk to take up the post as director of the Antiplague Institute. I feel it would be better to rectify my error now before committing myself further." Their reply contained the following: "On completion of your work you will be offered the opportunity of a transfer to scientific research in one or two years." An order from the minister followed, and there was nothing for me to do but comply. After handing over my department to Evgenia Bakhrakh, I departed for Irkutsk on May 15, 1957. I was then thirty-one years old.

For the transfer of administrative authority they sent along with me a commission consisting of Professors V. M. Tumanskij and V. N. Fedorov, who had always been kind to me, and M. G. Lokhov, a senior member of the All-Union Antiplague Mikrob Institute. These people were true plague experts. They wanted to familiarize themselves with the plague situation in Siberia, which was quiet at that time. They left Irkutsk soon after my arrival.

We traveled by train for nearly five days, but we were never bored! My fellow travelers were very interesting and enthusiastic people. I also saw the Ural Mountains and much of Siberia, which even then seemed to me a mysterious land of inexhaustible natural resources, a land of real heroes and of exile, a land like a bolt from the blue.

NOTES

1. Tambov oblast and Penza oblast are about 500 and 700 kilometers southeast of Moscow.

2. Collective farmers (*kolkhozniks*) had no passports till the 1960s and therefore were deprived of the freedom of migration.

3. Vladimir A. Haffkin was a medical doctor sent by the Russian government to fight plague in India at the beginning of the Third Pandemic in 1894. After that, till 1915 he worked in Bombay, where he established the antiplague station (later the institute, named after him).

4. Selman A. Waksman (1888–1973) was a leading authority on soil micro-biology. He received the 1952 Nobel Prize in medicine or physiology in recognition of his discovery of streptomycin, which became the first specific agent effective in the treatment of tuberculosis. Born in the Ukrainian region of Tsarist Russia, Waksman was educated in the United States and spent his entire professional career at Rutgers University, New Brunswick, New Jersey. (Ben Garrett)

5. Domaraskij's research interests at this period were the causative microbial agents for brucellosis, plague, and tularemia. These three plus the bacteria responsible for anthrax and glanders and the virus responsible for smallpox are the biological agents most frequently cited for their potential use in biological weapons. The Soviet system apparently encouraged Domaradskij's research interests because of the state's dual interest in these biological materials as public health menaces and as potential biological weapons. Similar dual interests applied in other nations, especially the United States, which had a robust biological weapons program during the period the author is describing (1950s). (Ben Garrett)

6. It is important to stress that Lysenko was supported not only by Stalin, but also by Khruschev.

7. Lysenko promised to develop new crops with higher yields quickly. Using a method called "vernalization," he claimed he could turn spring wheat into winter wheat, thus permitting a longer growing season, by soaking and chilling the seeds according to a prescribed pattern. Needless to say, it did not work, and the much of the blame for the food shortages that plagued Russia during the Stalin era can be laid at Lysenko's doorstep.

8. Thomas Hunt Morgan (1866–1945), geneticist and embryologist, was famed for experimental research with the fruit fly, *Drosophila*, which helped establish the chromosome theory of heredity. This theory conflicted with Lysenkoism, which rejected the concept that chromosomes played a specialized role in heredity. Morgan was honored as the 1933 Nobel Laureate in medicine or physiology for his work on heredity. (Ben Garrett)

9. Domaradskij's comments illustrate an element of graduate research that is largely without parallel in the West. Soviet candidates for advanced degrees had to tie their research to advances in political thought and doctrine. While such a tie might be straightforward in many social disciplines, it makes an awkward fit at best in the sciences. The author will illustrate this situation further when he describes how party politics replaced scientific achievement in assigning merit to research results. (Ben Garrett)

10. Katya, one of the sisters of my maternal grandmother, had to graduate from a medical department at a university in Switzerland.

11. In the USSR there was a permanent shortage of any reagents to help bring about chemical reactions. It was possible to get them but only with the ministry's help.

CHAPTER 4

IN SIBERIA

> Childlike, we trusted in fortune,
> In science, truth and men,
> And we faced up boldly
> To every storm.
>
> S. Y. Nadson (1862–1887)

The Antiplague Institute in Irkutsk was founded in 1934 by A. M. Skorodumov (born in 1888), whose memory is still zealously guarded by members of the institute. He was both the founder of the institute and a victim of Stalin's terror. As with many other members of the plague-control establishment, he met a tragic end. He was arrested in 1937 and perished in the dungeons of the security police.

From 1939 until his death N. A. Gaiskij (1884–1947), once head of the plague-control laboratory in the town of Furmanovo in the Ural Region, served as deputy director of the institute. Before his appointment he had successfully completed experiments begun before the war with B. Ya. Elbert to produce a live tularemia vaccine, which has helped to minimize tularemia morbidity among the people of this country.[1] Like Skorodumov, he suffered Stalinist persecution. In 1930 he was found guilty of crimes pursuant to Art. 58-11 of the Penal Code—in other words, of anti-Soviet propaganda—and was sentenced to five years in the camps, where he worked as a microbiologist in one of the specialized labor camps, called *sharashki*,[2] that frequently made state use of the work of highly skilled prisoners. After his release, incomprehensible as it may seem to Westerners, Gajskij continued working for the very system that had oppressed him. He worked as a bacteriologist, this time for a military laboratory. He is buried in the courtyard of the Irkutsk Institute.

Another remarkable man, I. S. Dudchenko, was linked to the Trans-Baikal area. Dudchenko founded the bacteriological laboratory in Chita, east of Lake Baikal, which officially opened on September 17, 1913. Under the laboratory's charter as confirmed by the Ministry of the Interior in Old Russia, its responsibilities included tackling scientific problems concerning plague, cholera, and other epidemic diseases. At the same time the laboratory was the central establishment for bacteriological research in the Primorsky Kraj[3] and the general government of Irkutsk. The life of Dudchenko was tragically cut short on the night of June 5, 1917. While preparing to travel to a plague area he and his seven assistants were savagely murdered by bandits who had broken into the laboratory building. Dudchenko was shot with a pistol, the others strangled with rope. Their bodies were thrown into a water-filled cellar.[4]

The Irkutsk Institute served a vast area stretching from the Altai to Sakhalin, fully living up to its title as the "Institute for Siberia and the Far East." The expanse between Altai and Sakhalin, the limits of the region, is approximately 5,500 to 6,000 kilometers and stretches over five time zones. Like the other plague-control institutes in the Soviet Union, the Irkutsk Institute oversaw a network of stations and sections. Stations—the larger administrative units—were situated in the worst plague regions. The most important of these stations were the plague-control establishments in the Trans-Baikal area—the region around the vast Lake Baikal, the greatest freshwater lake in the world, where outbreaks of plague, which began in the second half of the nineteenth century, continued until 1930. During the major epidemic of pneumonic plague there in 1910–11, academician D. K. Zabolotnij, a renowned Russian plague scientist, demonstrated the part played by tarabagan (a type of marmot, a large rodent) as one of the principal carriers of plague in nature. The tarabagan had been almost eradicated by the time of my arrival in Irkutsk. They had been the victims of an attempt by institute scientists to combat the infection by killing its carriers. Although repeated testing had been carried out, no cultures of plague microbe had been isolated for about ten years. But the plague-control establishment still paid close attention to this focus. For a long time, the institute attempted to combat plague by baiting tarabagan burrows with chloropicrin, a poisonous liquid.[5] Plague is transmitted from host to host by the bite of infected fleas, and can be transferred from rats or tarabagan or other rodent species to humans in this manner. Tarabagan are considered especially dangerous; it is thought by many Russian experts that the plague agent that comes from tarabagans is particularly virulent. Tarabagan hunters, for obvious reasons, are particularly vulnerable to this infection, and can introduce plague into human populations, as happened in the great epidemic of 1910–11. As we will see later, though, this attempt to combat plague by eradicating its carriers was not entirely successful.

There are no natural plague foci in the Russian Far East, but in 1921 a pneumonic plague epidemic erupted in the Primorsky Region and in the city of Vladivostok. Travelers on the Chinese Eastern railway from Manchuria had brought it into the area. The Chinese travelers were taken ill with pneumonic plague in Manchuria, probably from an outbreak among tarabagan hunters, and many thousands of people died. The fear of fresh outbreaks made it necessary to maintain plague-control establishments. Similar plague-control stations were built in Tuva and in the mountains of Altai because of the presence of potential natural foci, and the threat of rodent invasion from the active foci in Mongolia. The "active foci" are those where the plague microbe may be isolated at all times. Altai and Tuva were not active foci; the microbe was isolated from rodents from these regions relatively infrequently.

Besides plague, the Irkutsk Institute dealt with tularemia, for which natural foci are scattered throughout Siberia, and with brucellosis, which poses a serious risk to the public, even in the Far East. But in the mid-1950s the Plague Control Institute in Irkutsk was in a period of ideological crisis. There was no longer any direct risk of plague in the area, while the task of monitoring tularemia* and brucellosis† had now fallen to the epidemiological stations. Plans were being drawn up in Moscow to reconstitute the institute. This plan for reorganization was disturbing to the institute's workers: the plague foci of the region were quiescent at that time, and the workers could not find themselves new occupations, despite the existence in the region of many other infections.‡

The reorganization ultimately proved beneficial: eventually, the plague-control system in general and the Irkutsk Institute in particular were made responsible for the research management of the recently formed departments for highly dangerous infections at the levels of the republic, the regional, and the local epidemiological stations. In addition, the Irkutsk Institute became the leading establishment for leprosy (there was a leprosarium in Irkutsk, and there was also some leprosy in Yakutia, about two thousand kilometers away). Leprosy is a widely spread, very serious, chronic, contagious infection, known since antiquity, which involves progressive nerve and tissue destruction, disfiguring skin lumps and tumors, possible blindness, and the loss of extremities. The length of the incubation period is usually three to seven years, but may be as long as twenty to thirty years. Modern treatment of leprosy involves different sulfa drugs and the antibiotic rifampicinum, but no effective cure yet exists. The best preventive measure remains isolation of those already infected. But for them, the prognosis is poor. Nevertheless, the institute only performed this function for a relatively short time, and it was withdrawn virtually unnoticed.

---

*Tularemia is an animal disease carried by gnats that mainly affects water rats and small rodents (voles and mice). It is a widely distributed disease, even found in the United States. People can contact tularemia, which is not usually fatal even without treatment, though some strains may have a much higher virulence. It is not spread from person to person.

†Brucellosis is an animal disease that affects small- and big-horned herd animals (e.g., cows, sheep, goats). There is a good living vaccine against tularemia. Brucellosis proceeds often as a chronic disease. Both diseases are bacterial and are treated with antibiotics.

‡There are many different bacterial and viral infections in a huge territory of Siberia and the Russian Far East. Besides plague, tularemia, and brucellosis, there are also leprosy, pseudotuberculosis (an intestinal disease related to plague, but much milder), and various forms of insect-borne encephalitis (a disease that produces swelling in the brain).

Traditionally strong in the study of natural focal diseases, the Irkutsk Institute possessed a large staff of excellent parasitologists and zoologists, headed by some outstanding personalities and distinguished specialists such as I. F. Zhovtyj (parasitology) and the native Siberian N. V. Nekipelov (zoology and epizoology). Both were well known for their work in plague research. But in the fields of modern microbiology and general epidemiology, the institute was much weaker than other plague-control centers.

By the time I was transferred to Irkutsk in 1957, the institute had begun to undergo dispersal and fragmentation. There was no unanimity between research collaborators; the former director could not put the place in order and so he lost respect and prestige in the eyes of his staff. I was the only one in the institute with a doctorate, so a great many of its members made me welcome, promising all kinds of help and support for the restoring of order. (Unlike Saratov, where my doctorate at so young an age had been resented, the Irkutsk staff seemed to welcome my credentials, youth, and energy.) Many of them were expecting me to bring some fresh ideas and stimulation into their scientific work. With the passion of youth, and spurred on by ambition, I plunged into my work.

The principal role of the institute was to combat any epidemics in this vast area of the country; my first task was to find a widely experienced researcher to head the Epidemiology Section. After much trial and error, and by a fortunate chance, V. A. Kraminskiy turned up. He had previously been head of the Epidemic Control section for the Far Eastern Military District.[6] I had met him first at a conference in Irkutsk, and shortly thereafter he left the army.[7] With my deputy I. F. Zhovtyj as parasitologist, N. V. Nekipelov as zoologist, and V. A. Kraminskij as epidemiologist, I had a powerful trio with whom to work. With such specialists I could solve problems I had never encountered before. And they never once let me down! Our trio was supported by some magnificent specialists in tularemia (M. I. Antsiferov), brucellosis (A. F. Pinigin), and also by L. A. Timofeeva, who had been head of a microbiological laboratory.

We immediately began to familiarize ourselves with our "periphery," which brought me enormous pleasure. I criss-crossed Siberia and the Far East, turning up in places I had never dreamed of visiting. But this was no recreational getaway. It meant some arduous journeys packed with meetings and discussions. We discovered natural plague foci in Tuva and the mountain Altai, which some scientists had doubted existed before. The antiplague station in Tuva was a very low-level one; it could only manage to control the sanitation in this region. Discovery of the natural plague foci, after appointment of a new specialist in plague, exceeded the capabilities of this station. These foci were discovered by the combined efforts of the Tuva Station and the Irkutsk Institute. The plague carriers proved to be a local species of rodents.

These excursions were not confined merely to Siberia and the Soviet Far East. Only a few months after I moved to Irkutsk, I visited China, not by the conventional route, via Peking, but via Urumchi in the northwest. In company with two well-known plague experts, B. K. Fenyuk (from the Saratov Mikrob Institute) and V. S. Petrov (from the Central Asian Anti-plague Institute at Alma-Ata), we toured all the natural plague foci in China, including those in its southeast region. In many Chinese towns we delivered reports and lectures that were always well attended, and some of our material was published in Chinese. Our cooperation with Chinese scientists was very much in the interest of both countries. After the Chinese revolution in 1947, the national economy was on the decline, and the USSR helped it, in particular, to recover good standards of public health. Our cooperation concerned the fight against plague as well, since plague, as we see throughout history, knows no national boundaries.

This tour was extremely important, since up to that time we had known about Chinese plague research only second hand. For more than twenty years China had been closed to us. A large part of China, and Manchuria, a region most relevant for plague research, had been occupied by Japan, our traditional enemy in eastern Siberia and the Far East, from the early thirties to the end of the Second World War. My colleagues benefited from this expedition more than I did, since I was still a neophyte in epizootology, the study of infectious diseases of animals. Nevertheless I also learned a great deal, and the trip left me with indelible impressions. We caught sight of the cloud-piercing peaks of the Nanshan and Lake Kuko Nor, the Jungarian wastes, the rivers Yangzi and Huanghe, endless ricefields and sugarcane plantations, mandarin-orange groves, and great cities. We had the good fortune to see Hangzhou ("in heaven—Paradise, and in Earth—Hangzhou") and to walk on the Great Wall!

This was my first trip abroad. Then I was also twice able to visit Mongolia, whose plague-control service was regularly supplied with methodological assistance from our institute. The Irkutsk Institute is not far from Manchuria and Mongolia, sites of active natural plague foci. Our experts in plague, therefore, worked in Mongolia on an almost permanent basis. I particularly recall our second visit: we had come to investigate suspected pulmonary plague in Ulan Bator, the Mongolian capital, which is located almost on the Russian/Mongolian Chinese border. I was invited to meet Vyacheslav M. Molotov (of "Molotov cocktail" fame), once Stalin's right hand, who by then had been relegated to an ambassadorship. He was interested in the plague situation in that country and I presented him with a brief account. But, more important, my colleagues and I were granted the rare opportunity to examine a number of the main natural plague foci in Mongolia, among the large colonies of tarabagans native to the region. Plague outbreaks among human beings in Mongolia, as we have seen, are connected with the hunting of marmots; they are the main source of the plague microbe in the area. We first flew to Bayan-Hangor and then went by car through the wilderness of the Gobi-Altai to the Altai Mountains border.

The second trip did not go off without some adventures. I had been invited by Mongolian authorities to a conference on plague. Someone had made a mistake with our papers in Irkutsk, so that I was held up by a Soviet border functionary in Kyakhta as someone "attempting illegally to cross the frontier"—and was turned back. I had to travel all the way back to Irkutsk, and this time take a plane directly to Ulan Bator. As a memory of those journeys I still have a wooden kumiss bowl—a special vessel for the Mongolian drink kumiss, soured mare's milk—decorated with silver. The Mongolian minster of health presented this to me.

Another part of my job in Irkutsk was to organize new laboratories for biochemistry and pathophysiology, the part of physiology devoted to the study of the mechanisms of disease. But for my own principal subject of research I elected to study the mechanism of plague toxicity (how plague poisons the body), a topic that had been scarcely investigated at the time. The outcome was my first monograph on the pathogenesis of plague,[8] published after I left Irkutsk. I also produced two surveys, at least one of which has retained its importance.[9] Since then I have addressed the problem of toxins several times, but from a general biological standpoint, which has been developed in studies by other authors. Many experts now agree that, in the agents of numerous infections, the role of many toxins is to protect their producers from aggressive action by other microorganisms that inhabit the environment. Toxins are poisonous substances; in particular, they are produced by bacteria. They play an important role in how infections develop. Sometimes they are directly responsible for the deaths of human beings and animals (e.g., plague, tetanus, botulism, cholera toxins). The plague toxin, its mode of action on organisms, and its other properties have not been fully explored even now. (As we will see in later chapters, the Soviet bioweapons program "Factor," which I helped create, attempted to genetically engineer the production of toxins by viruses or bacteria as a way to make those engineered strains even more virulent. My interest in this issue eventually produced conflicts with the heads of some of the bioweapons institutes, who did not have the same interest in the properties of toxins as I did.)

In addition to plague pathogenesis, my attention was also drawn to the subject of crossover immunity, the study of which served to stimulate me to write what is still the only monograph, at least in Russia, on "Problems of Crossover Immunity."[10] "Crossover immunity" is the ability of one microbe to create immunity against an infection caused by another species. I have also written a long article under the same title published in 1993.[11] I continue to feel that this line of research holds considerable promise in the fight against infectious disease; the capabilities of crossover immunity should not be ignored. The English physician Edward Jenner, who used vaccination with cowpox to combat smallpox in humans, demonstrated these capabilities as long ago as the late eighteenth century.

On the appearance of V. A. Kraminskij, the institute began to concern itself with virology as well, which was not in the program of the other plague-control establishments apart from the Antiplague Institute in Stavropol. Our interest in virology developed mainly as an initiative by Kraminskij's wife. I supported her initiative because tick-borne encephalitis in eastern Siberia as well as mosquito-borne encephalitis in the Russian Far East were severe and widespread problems.

One other innovation was the production of a cholera endotoxin, used for vaccination against cholera, which, owing to the vast amounts of alcohol needed for its production and purification, attracted the attention of the whole domestic staff of the institute! Some of them even tried to consume spent alcohol, with its inevitable contaminants after laboratory processing, so we were forced to drain it off into containers sunk into the ground.

Governmental experiments were carried out under my supervision at the institute regarding a new live plague vaccine consisting of two strains (the "bivalent" vaccine 1–17), one of which (strain 17) had earlier been created in the institute. This vaccine was supposed to afford protection from plague agents of different origins: one strain had been isolated in Siberia from marmots and the other one from field voles in Turkmenistan. However, the production of this vaccine was continued for only a few years, because production of the "bivalent" vaccine was too labor intensive, and the strains were forever dissociating or growing more variable and hence unreliable in the process. It was also found later that the new vaccine possessed no marked advantage over the live EV vaccine, an avirulent strain of plague originally isolated in Madagascar by Girard and Robic,[12] which is still used today as a vaccine in Russia and some other countries, though not in the United States.

My work was presided over by then Deputy Minister of Health Viktor M. Zhdanov, to whom I was obligated for my appointment at Irkutsk, and whom I was to know later in my life in quite another capacity.

On my initiative, furthermore, our institute was one of the first in the country to develop a technology of dry nutrient media* for the cultivation of bacteria, which, incidentally, it is still producing today.

Over a number of years, and despite the best efforts of members of the Chita plague-control station, it had proved impossible to isolate a culture of plague microbes in the Trans-Baikal focus. On this evidence people began to speak of the area's being cleared of plague. To test this, we allowed the highly plague-sensitive tarabagan population in the 62,000-hectare area around the Torei Lakes to recover. Four years later the cultures had reappeared! I therefore became convinced that it is scarcely possible to eliminate the natural foci of infectious disease without causing unjustifiable damage to natural species, although this meant "cutting off the branch" on which the plague-control establishment had grown. The baiting of tarabagan burrows ceased.

As we now know, many years later, the attempt to kill off entire host populations of rodents is not by any means the most effective manner of plague control. This has been proved by extensive experience in the USSR and other countries, including the United States, where attempts to destroy whole prairie dog populations have not eliminated the disease. The eradication of various rodents (not only the tarabagan) does not guarantee the disappearance of plague from a region. Furthermore, outbreaks of plague connected with rodents are relatively infrequent occurrences and may be controlled by antibiotics. I believe that the former point of view, that the eradication of rodents as the principal means of fighting plague, was erroneous.

There are, furthermore, many other reservoirs of animal infection—anthrax, for example, which is found around the world among cattle, sheep, and camels, or brucellosis, which is also a livestock disease. This does not mean that we combat anthrax or brucellosis by eliminating our cattle.

But from a scientific point of view this reawakening of a plague focus was of enormous interest, because it confronted us with the persistent question of what happens to plague bacteria in periods between out-

---

*Nutrient media are broth or jell used to grow colonies of bacteria. Dry nutrient media have advantages over traditional preparations because they are more easily stored and less likely to spoil.

breaks. Since the cultures isolated in Gulzhunga, the area around the Torei lakes, were found to be weakly virulent—much less deadly than most other plague strains—V. A. Kraminskij and I developed a hypothesis suggesting that such less virulent strains can play a role in the cycling of plague in rodent populations, and maintaining the strain in nature. Very deadly plague strains, in contrast, can burn out, killing or immunizing their entire host population, so that the strain vanishes from the region until it is imported again.

Although the idea that lower virulence may evolve to maintain strains in nature has been restated by Frank Burnet[13] with respect to some other microorganisms, and is shared by many foreign scientists, a number of objections have been raised by Russian plague experts. But the validity of our hypothesis in the case of plague is supported by a great many fresh facts that answer the objections. The orthodox position on epizootic processes in plague maintained that strains with reduced virulence could not circulate in nature. But in recent years data has accumulated regarding the heterogeneity of virulence in populations of the plague microbe. There is also evidence that different plague phenotypes produce different degrees of virulence; furthermore, strains isolated from field voles have *all* been shown to be weakly virulent. Finally, it is now apparent that the virulence of *Y. pestis* plays an ambiguous role in epizootic processes. All of this makes it possible not only to refute these orthodox objections, but also to further develop our own hypothesis. Incidentally, the World Health Organization (WHO) Expert Committee on Plague has essentially supported our position.[14]

I am particularly proud of the published work that emerged from my activities in Irkutsk, which was always supported by my colleagues, and principally by I. F. Zhovtyj, V. A. Kraminskij, and N. V. Nekipelov. Even in the recent past, gaining admission into the scientific journals was seen as almost a victory, especially for those working at the fringes of science. But the publication of the *Bulletin of Microbiology* by the All-Union Antiplague Mikrob Institute had already ended at volume 19, before the war, while the issue of works from the other institutes was a great event. Throughout my period of work in Saratov I was able to get only one article into a journal! There was an absence of publications after 1956, when a new rule appeared from the State High Certification Commission (VAK), preventing the defense of dissertations for several years. (I had managed to submit my defense only one month before.)

It was not even a matter of money (which was always a factor), but of permanent weakness in the printing establishment, which was engaged in printing only ideological literature. Furthermore, we were hampered by the constraints of censorship. In the USSR the publication of true statistical data on almost any infection was prohibited. The Soviet government loudly proclaimed that "a number of infectious diseases have been eradicated in the Soviet Union" and that "others will likewise be eliminated in future years." All cases of infection with plague and cholera after 1938 were declared secret: this is how plague and cholera were "eliminated" in the USSR.

Publishing work on the causal agents of such dangerous infections was still permitted. It was therefore natural for all of us to want to begin to publish again, and with some difficulty I managed it. In principle, institutes had the right to publish proceedings and reports. Apart from financial constraints, there were other difficulties: from the beginning of the USSR to its demise in 1991, censorship ruled! Censors could reject any publication (even scientific ones) without any explanation. Nevertheless, several volumes of "Reports of the Irkutsk Antiplague Institute" came out one after the other. In the first I published all of the articles that formed the basis of my doctoral work. Thus, the first studies in the biochemistry of the plague microbe saw the light of day. Afterward the situation became more difficult (funds for publishing began to be cut back) but we found a way out of this, and started to issue "Papers of the Irkutsk Antiplague Institute" (which were shorter pieces, and therefore less costly to publish).

Besides the publishing work, I paid a lot of attention to organizing various conferences both at the plague-control stations and in Irkutsk itself, with invitations to workers in departments of highly dangerous infections in the SES[15] for Siberia and the Russian Far East. Reports on these conferences appeared in both the local and the national press.[16]

Meanwhile, normal living conditions remained just a dream for the great majority of people, and many members of the institute were still housed in construction huts left over from the building of the institute in the 1930s. Even the leading specialists were living in extremely poor conditions, which affected their work in various ways. This was the "Soviet mode of life": scientists and specialists, as well as everyone else, did not then imagine any other existence! While a movement was afoot during Khrushchev's time (1953–1964) to shift people from housing unfit for normal human habitation to new places, I am reminded of his emphatic assertion that our generation would be able to *live* under Communism!

I tried to change that for the better, even if only for a few. Immediately after my appointment as director I began to push for the construction of housing. The first building, a small wooden house containing eight flats, was built by the "private" method,[17] which caused a lot of trouble with the ministry owing to my sundry financial blunders. The second building, brick, with forty-eight flats, was built not far from the institute and was quickly finished. It was built mainly for leading scientists and specialists. Both houses were built by financial support of the Ministry of Health.

I also had some trouble with this second building, but trouble of a different sort. When the building was ready for occupation and the municipal council had confirmed the list of those who were to move in—those who were in special need of quarters or who were leading members of the institute who in the event of emergency travel must always be "on call"— part of the building was informally occupied by people who were not on the lists. One cold winter's night I was rung up from the institute and informed of this "emergency situation." I immediately set about ringing up the council authorities, the municipal Party Committee, and even the public prosecutor, but from all sides came the reply: "That's your business, we can't interfere, but unless that building is cleared by the morning your Party card is on the line." (At that time in Irkutsk there had been some illegal seizures of buildings by people at the end of their tether, who showered the chairman of a routine Party conference with telegrams, and the local authorities had to deal with a spate of unpleasantness.) I had to race to the institute and collect some Communists to clear out the building (no one but Communists would cooperate in throwing the squatters out of their new quarters). When I got to the house in utter darkness with a small team of workers, I was met by an ugly crowd of people defending those who had moved in. Followed by shouting and threats, we burst into the building and began clearing the occupied flats by force. Some flats even contained furniture and flowers in the windows. In one flat someone went at us with an ax. But by morning it was all over.

In general and without false modesty I have to say that under my direction the institute had been raised to a new and higher level. I formed a capable working collective by assembling a superior group of researchers and experts. The scientists at the institute, through my initiative, published a wide range of work. I was able to improve the lives of my colleagues by the building of suitable housing for them, and I myself had the opportunity to carry out significant investigations of cross-immunity and of the pathogenic factors of the plague microbe, focusing in particular on the virulence factor fibrinolysin.

Peace prevailed throughout the institute and the whole atmosphere was changed. This is not only my opinion but also that of the people who were working hand-in-hand with me. To my knowledge I never put a wrench in anyone's works and I always upheld any sound initiative. Of course there were some hitches, but there always are.

I was quite surprised therefore to learn from a book,[18] of which one of the authors, E. P. Golubinskij, was a student of mine, and is now the director of the institute, that my chief merit is seen to have been merely the organization of the biochemical laboratory! This seems still more peculiar considering that it was I to whom Golubinskij largely owes his promotion. Anyway, this is just one more confirmation of the old saying that "No good deed goes unpunished." It is not the first time I have encountered it in life.

The move to Irkutsk radically altered my life. There for the first time I had my own flat. Though my first flat, in a very old house, completely lacked any amenities, including water (which had to be brought in by water carrier), and had only coal-fired heating, I felt like a human being. It was in Irkutsk, too, that I finally settled my personal life. I married again: my marriage provoked a considerable reaction (including some anonymous letters). This was mainly because my wife was an actress. Even my parents-in-law predicted serious troubles for us in our future life because of the incompatibility of our professions, and to some extent they were right. But everything was fine in the end. Irkutsk became the birthplace of our two daughters, and my wife was a great success in one of the oldest theaters in Siberia.

We had many friends and we lived very happily. I think that, had I remained in Saratov, despite my doctorate I would have led a miserable existence. My salary was much higher as head of the Irkutsk Institute,[19] and I eventually obtained a very good flat, in contrast to my waterless first residence in the city and the cramped quarters I had in Saratov. In Irkutsk, also, my life had a much broader scope. I made the acquaintance of several very eminent people (actors, musicians, and scientists), including those who had moved from Moscow. While living in Irkutsk I also had the chance to see the world. Besides China and Mongolia I traveled to eleven other countries in Europe and Asia.

In Irkutsk I was elected a deputy of the Municipal Council of Workers. I derived no joy from the fact, especially since it only made me a member of the Establishment, and was not an acknowledgment of any particular merit. At the same time or a little earlier my great friend Prof. V. A. Pertsik became a Deputy of the Municipal Council. He was a well-known jurist whose main specialty was "Soviet construction," i.e., the drafting of various standards documents, in particular, designs for the improvement of local "self-government." I was rather lax in my duties as a deputy, feeling convinced that I could hardly bring about any change. Pertsik and I argued frequently about the role of deputies and the institution of councils in general, and each time each of us stuck to his own opinion.

During my time in Irkutsk, the mass exodus of Russians from China during the "Khrushchev thaw" in the later fifties had already begun. Many of these people had fled the establishment of Soviet power in Siberia and settled in China. Worsening relations between the two countries forced their return. (Many of these people, after their return, were subject to charges of espionage against the Soviet Union.)

Among these repatriated Russians there was one old woman who became our housekeeper in our new flat. Until then she had spent the whole of her life in service with the manager of the East China Railway, and she was quite unable to get used to the new habits and customs. Every time she watched how we amused ourselves—we sang, danced, and played the fool as do all young people—she would shake her head reproachfully and grumble, "Look at them, members of the municipality! Did you ever see anything like it!"

All in all, this was one of the happiest chapters of my life, and I saw it draw to a close with apprehension and with regret. But I would have to leave Irkutsk much as I had left Saratov. It was not easy, as I discovered on several occasions, to buck the System.

NOTES

1. As a rule, the live vaccine protects against tularemia much better than other forms of vaccine.

2. The *sharashki* were the Inner Circles about which Alexandr Solzhenitsyn has written. (Ben Garrett)

3. Primorsky Kraj is in the Far East of the USSR/Russia.

4. In the unsettled days of the Revolution, chaos and banditry extended into Siberia. The tragic death of Dudchenko was a direct result of the widespread disruption of Russian civil society.

5. A cotton wool moistened with chloropicrin was left as bait in tarabagan burrows.

6. The Far Eastern Military District is located on the Primorskij and Khabarovsk Krais and a part of Eastern Siberia (Amur oblast, after the Amur River). Russia is divided into eighty or so oblasts (or regions) and six krais (Stavropolsky, Krasnodarsky, Altaisky, Krasnoyarsky, Khabarobsky, and Primorsky). Krais are the same as oblasts, but have in their structures autonomous oblasts (regions) or republiks, and are near a frontier. Usually, but not always, oblast are named after their main city (Moscow and Moscow oblast and so on). Two exceptions are St. Peterburg and Leningradsky oblast.

7. The ministry hires as a rule only the heads of institutes. Other scientific

researchers are elected usually by the academic council of the institute or are hired by its head (in extraordinary cases).

8. I. V. Domaradskij, *Ocherki Patogeneza Chumy* (Sketches of Plague Pathogenesis) (Moscow: Meditsina, 1966).

9. One of them was devoted to a mechanism of toxin action, and the other to the antigenic properties of enzymes.

10. I. V. Domaradskij, *Problemy perekrestogo immuniteta* (Problems of Crossover Immunity) (Moscow: Meditsina, 1973).

11. "Crossover-Immunity: Perspectives and Hypotheses" (in Russian), *Immunology*, no. 3 (1993): 6–9.

12. G. Girard, L'immunity dans l'infection pesteuse (Immunity in Plague Infection). Acquisitions apporte par 30 annees de travaux sur la source de Pasteurella pestis EV (Gerard et Robic). *Biologie Medicale* 52 (1963): 631–731.

13. Sir (Frank) Macfarlane Burnet (1899–1985) was an Australian physician and virologist. Burnet received the 1960 Nobel Prize in medicine or physiology for the discovery of acquired immunological tolerance to tissue transplants. He also isolated the causative agent of Q fever, and the species bears his name in his honor (*Rickettsia burneti*). (Ben Garrett)

14. WHO Expert Committee on Plague, Fourth Report (Geneva: World Health Organization, 1970).

15. Sanitation and infection-prevention stations, located in cities, villages, and towns throughout the Soviet Union.

16. See, e.g., *Meditsinskaya Gazetaö*, no. 3 (January 9, 1959).

17. That is to say, not through any state building organization.

18. E. P. Golubitskij, I. F. Zhovtij, and L. B. Lemesheva, *About Plague in Siberia* (Irkutsk: 1987).

19. Western readers may be less familiar with the practice, common in the Soviet Union and perhaps found in certain other societies, to reward a scientist with a better residence or with other, domestic advantages based solely on one's position (in the author's case, that of institute head). Nonmonetary rewards such as enhanced assigned housing, vehicles, and access to certain otherwise off-limits stores, were routine and expected for workers in all disciplines in the Soviet Union, including scientists. (Ben Garrett)

CHAPTER 5

Go thine own way, and let them say what they will.

Dante

I was happy in Irkutsk: the institute had been placed on a sound footing; my wife's acting career was flourishing; we had a good living situation and many friends. We had no desire to leave and begin a new life elsewhere. But larger forces were soon to intervene, which would set my life and career on an entirely different trajectory. In 1963 or 1964 the plague-control establishment underwent a thoroughgoing reorganization. The Soviet government had long had its eye on developing an ambitious biological weapons program; it was also well aware that such programs were advancing in the West. Defense against biological weapons, the so-called Problem No. 5, became the main focus of the Antiplague Institute in Rostov-on-Don, which was deprived at the same time of its traditional purpose, namely the study of plague foci.

This growing concern with biodefense coincided with an attempt to resuscitate the all but defunct study of genetics in the country: after the depredations of Lysenkoism and the attempt to ban the study of Mendelian genetics, knowledge of modern genetics and molecular biology was at long last beginning to creep into Soviet Russia from the outside world. It became increasingly clear that the solution of many bio-logical-defense problems called for a knowledge of modern biochemistry, and the institute at Rostov, while it had a biochemistry lab, was at that time only working on immunochemistry. So the powers that be offered me the chance to move to Rostov-on-Don as the head of the institute there. Superb though the offer was, I was in some ways reluctant to accept it—as they say, "Good things never turn out well."

I faced this new offer with deep ambivalence. On the one hand, it was a pity to leave an institute where I had worked successfully and happily for seven years, and Irkutsk, a city where my wife and I were happy. Fur-thermore, the upheaval we would have to endure while moving was a dismal one: in the USSR any move from one city to other was always attended by considerable difficulty. On the other hand, the prospect of a new horizon was alluring. The reorganization of the Rostov institute promised rich scientific possibilities. Also, Rostov is nearer to Saratov, where my mother and mother-in-law lived, as well as to Moscow with its well-known institutes and libraries.

Still, I tried to turn down the move. But in the end I had to comply: in the usual manner of the time, I said, "What must be, must be."

At that time a serious situation had developed at Rostov. The institute was being ripped apart by infighting, as a result of which its operations were almost completely paralyzed. When the discussions began con-cerning my transfer, echoes of this infighting—in the form of anonymous letters—resounded all the way to Irkutsk; they consisted both of threats and of invitations to me to bring about some order at Rostov. The main ground for these hostilities was that a number of leading members of the institute, in particular Prof. M. I. Levi, a gifted scientist, bitterly resented the decision by the Soviet Ministry of Health to change the direction of the Rostov Institute's activities from the study of plague foci to "Problem No. 5" issues of biological defense. These scientists had done a great deal of work in epizoology and in developing new methods of diagnosing plague. These advances put their research into rapid plague diagnosis and the study of rodent epidemics on a stronger footing and altered their entire method for keeping track of the natural plague foci. These Rostov researchers had worked out a new method of plague diagnosis based on the ability of human or animal erythrocytes (red blood cells) covered with antibodies to leave sediment over plague microbes. The method is very suitable for searching for plague in animal foci in nature. It may be used instead of the usual long procedure of microbe isolation for the diagnosis of many other infections.

Naturally these scientists did not want to abandon their plague-control work to switch to biological defense. Furthermore, some zoologists, parasitologists, and even epidemiologists at the institute were either unable or unwilling to completely change the direction of their research as required by the new tasks that the Ministry of Health had devised.

I was therefore greeted with fixed bayonets, so to speak. Apart from a few people, the scientists at Rostov regarded me as well nigh a personal enemy. It was like being in a fortress under siege. My invidious position was aggravated by a certain arrogant and supercilious attitude on the part of the Regional Committee of the Communist Party. During my first meeting with the secretary responsible for science, he said to me, "This is Rostov, not Irkutsk!"

It was especially unsettling to experience the outward fawning, though at bottom hostile, attitude of V. L. Pustovalov, the head of the bio-chemical laboratory, where the equipment was far superior to what I had had in Irkutsk. He seemed to me a very strange and narrow-minded person who brushed off any interference in his affairs. He had good reason to resent me, since his laboratory was to be reorganized to become a department under my authority. He also resented the direction of my research, the study of metabolism, in which he was not very well versed, and which he believed to be of lesser scientific importance. Despite his total lack of interest in my work, he still considered me his rival.

Pustovalov also could not understand that his attempt to create an effective "chemical" plague vaccine was hopeless, as not nearly enough of the antigenic structure of plague had been studied at the time. In other words, the structural factors he would have had to place into his chemical vaccine to produce an immune response were still undiscovered, but Pustovalov did not want to stop working on his project. As a result, his work was pointless, and he was, in effect, simply "marking time." The hostile nature of our relationship never changed. It was soon evident that there was no point in being conciliatory, even if I had wished to be.

This meant that even the matter of my accommodation was unresolved. At first I had no flat in Rostov but was forced to live in a hotel, which was very expensive. Eventually I had to move to the isolation unit of the Institute, where I lived until after a full five months I obtained a flat and my family could leave Irkutsk to join me. The flat was in a nasty district. Although it was a three-room flat as promised, it was small, uncomfortable, on the ground floor of a nine-story apartment building, and distinguished mostly for some very inconvenient planning. I hated to bring my family to such a dismal place. It looked out on a fence surrounding a factory. It was there that my wife arrived with our daughters, the younger of whom was only seven months old. And we would have had to live there until Doomsday unless something was to turn up.

But almost immediately after my arrival I managed to have completed the construction of a house owned by the institute just when the regional committee needed a single-room flat. And with the permission of the Soviet Ministry of Health, in exchange for this flat in the institute's building (even though it truly was not fit for human habitation) and the dismal one in which we were already living, the committee gave me a splendid flat in a very good building in the center of Rostov. For this we were obliged to the second Secretary of the Committee, V. F. Mazovka, a very decent person— unlike most party leaders—who apparently liked me, and whose grandson was in the same class as my daughter at school. He was kind enough to help us. At least our housing problem was out of the way.

My wife had had to give up her work as an actress in Irkutsk, a city with a fine theater and an old theatrical tradition, and come to Rostov, where no immediate prospects awaited her. Finding work for her in a theater took longer, but we managed it after much trouble ("This is Rostov, not Irkutsk!").

But other difficulties remained for us: I found that, even after many months, hostilities toward the new direction of our research persisted at the institute. I had to resort to some unpopular measures: as director I had the right to dismiss members of the scientific staff. I fired a number of employees and threatened a few others with sacking. After that, the others began to be "house-trained."

Ultimately everything came under control, paradoxical as it may sound, when a cholera epidemic broke out in Karakalpakiya, Uzbekistan. We had to send in an epidemic-control team. This was a novelty for the Rostov staff, since everyone there had been taught to think that we had wiped out cholera. Recognition of the danger and seriousness of the situation united people and made them forget their petty squabbles. Working amicably together we collected heavy equipment and loaded it onto the aircraft. I also flew with them to Nukus, the capital of Karakalpakiya, an autonomous republic of Uzbekistan. It is located on the Amu-Dariya River near the Aral Sea.

Besides the Rostov team, detachments from Moscow, Irkutsk, and other plague-control establishments assembled by the Ministry of Health were also present in Karakalpakiya, which had no cholera experts of its own. Each had its own area of work. For example, we were attached to the temporary hospital in Nukus itself. Authority was granted to A. I. Burnazyan, the deputy minister of health of the USSR, and head of the Third Main Directorate.[1] Every evening Burnazyan assembled the cholera-control staff, which consisted of I. I. Rogozin, V. Kraminskij, and several other well-known epidemiologists. Zhukov-Verezhnikov, a man of vast experience who was formerly both deputy minister of health of the USSR and a vice-president of the Academy of Medical Sciences, was appointed the scientific director.

Cholera is a rapidly spreading, serious acute intestinal disease characterized by sudden onset, profuse watery stool, vomiting, rapid dehydration, acidosis, and circulatory collapse. Death may occur within a few hours of onset. Mortality in untreated cases may exceed 50 percent. (Mild or even completely symptomless cases also occur during epidemics, which may be much more frequent than clinically recognized cases.) The diagnosis is confirmed by culturing cholera vibrios—a comma-shaped, mobile bacterium—from feces, or by demonstrating a significant rise in titer (concentration) of antibodies in acute and convalescent sera.

Strains of cholera include the classical and El Tor biotypes. The diseases caused by these variants are clinically almost indistinguishable, though El Tor tends to be milder. In any single epidemic one particular variant tends to be dominant.[2] Contamination of water supplies with cholera vibrios may quickly lead to a rampant and dangerous epidemic.

When we arrived, we found that the entire region of Karakalpakiya had been cut off from the outside world by army units enforcing a quarantine. There was scarcely any communication with the outside, and it was virtually impossible to telephone anywhere. Throughout this period I managed to call Rostov only a couple of times through some friends in one of the railway public-health units (who had their own communication system). It was possible to travel out of Karakalpakiya on a special permit only after six days of observation with a thrice-repeated bacteriological examination to make certain one was not a carrier of *Vibrio cholerae*.

The spread of cholera throughout Karakalpakiya and other parts of Uzbekistan was largely due to the delayed diagnosis of the early cases. Doctors were unfamiliar with cholera, and its occurrence within the Soviet Union had been ruled out (like plague, it had been "liquidated"). Another reason to keep quiet was that the government did not want to start a panic in the land from the dreaded word "cholera." So someone came up with a "legend": people were suffering not from infectious disease, but from defoliant poisoning. It was impossible to promulgate the legend for long though: defoliant poisoning was never so widespread as the cholera epidemic in Karakalpakiya.

So a new "legend" was created: the cholera had not originated in the USSR, but had come in from Afghanistan! At that time, we had a virtually open border with Afghanistan, so it was possible to maintain the fiction of the outbreak's Afghan origin. This explanation prevailed for a long time.[3] One job for the antiepidemic staff, indeed, was to investigate precisely any possible pathways for such a transference. But acknowledgment of the presence in the Soviet Union of indigenous foci of cholera occurred much later. It is not generally accepted even now: for example, the cholera epidemic in Daghestan in 1994—the first serious cholera epidemic in recent years in Russia—is still considered by authorities as an introduction from outside. The actual number of people suffering and dying from cholera in Karakalpakia was kept secret even from us, let alone from the public. This information was considered classified. It was known only to a narrow circle of people (e.g., A. Burnazyan, N. Zhukov-Verezhnikov, and their closest collaborators), as well as to doctors who actually worked at the hospitals. Of course, exact data was passed on to Moscow, too. But I was then not among the elite, so I was not made privy to exact information. Communication between various units (different kinds of hospitals and laboratories) was prohibited in an effort to prevent the leaking of the true information. This necessarily hampered the whole epidemic-control effort.

At first many idiocies were perpetrated, such as sprinkling the streets with creosote (a very toxic liquid containing phenol: since human beings carry cholera and contaminate others through fecal/oral contact, sprinkling the streets is useless!) or banning the posting of letters (also futile), and the key means of prevention was tetracycline, which they were dishing out in handfuls! I was not directly connected to the "holiest of holies"—the cholera hospital, where only the "elite" or those doctors directly involved in caring for the sick were allowed to go—and I don't know whether they were treating the sick with special saline solutions to prevent dehydration and the disruption of electrolyte balances, a principal cause of death from cholera. But I myself heard Zhukov-Verezhnikov argue with one of my colleagues, the pathophysiologist Avrorov, casting doubt on the effectiveness of the "Phillips" solution, a cholera treatment developed in the West, and calling the recommendations for their use, which had won worldwide recognition, "intrigues of SEATO,"[4] and "attempts at misinformation." Zhukov-Verezhnikov, spouting off the perennial Soviet dogmas, insisted that this treatment, which had saved innumerable lives in the West, was worthless, and that "imperialists purposely mislead us."

The laboratory had a colossal workload. My own, for example, performed one thousand or more analyses per day. For the microbiological diagnostics we generally used a classic method: we isolated cholera vibrios from human stool on liquid and solid nutrient media. To run one analysis, we needed approximately ten test tubes and several petri dishes. Owing to a water shortage the number of used petri dishes overflowed the baths in which they were washed. This was very dangerous work: the water teemed with vibrios, and we could easily have been infected. The water shortage was due to drought, a common occurrence in Karakalpa-kiya. The epidemic was made even worse by the primitive sewer system in Nukus. In spite of the fierce heat there was nowhere to cool off. (I finally adapted some toilet bowls for this purpose; you just pulled the plug and it sprayed out water, giving us a shower!) Everybody worked with next to no sleep, but with great enthusiasm. It is difficult now to single out any particular individual, although I would hand the palm to S. K. Ras-sudov, thanks to whose persistence and energy we managed to observe elementary safety while working with the cholera cultures.

Despite these difficulties, I myself tried to combine routine work with scientific research. It was not easy under such conditions, but interesting for the young people on fellowships who accompanied us. Without their help I could do nothing. After the Nukus epidemic, some of them pre-pared candidate theses based on their research. In particular we researched new methods of cholera diagnosis, simpler and more rapid than the "classic method" described above. We also searched for factors of vibrio El-Tor that might help to distinguish it from classic Vibrio cholerae and other bacteria. But I needed reagents and equipment, as in Nukus and even in Tashkent (a capital of Uzbekistan) to get them was impossible. There was a shortage of nutrient media and even such simple reagents as alcohol or dyes (for microbiological purposes), as well as shortages of basic equipment like petri dishes or glass pipettes.

To acquire nutrient media (agar-agar, or meat broth), which was absolutely essential to our investigations since it is the substance on which bacteria grow, we turned to our own laboratory at Rostov. These nutrients were delivered in liquid form by air in a glass container, which meant that despite the huge demand they sometimes spoiled in transit. (If nutrient media become contaminated by an overgrowth of bacteria, they obviously cannot be used again to grow only the bacteria you want!) Liquid nutrient media, which are by definition an environment friendly to bacteria, are especially susceptible to spoilage. Dry media are less easily spoiled, but we had none. On one splendid day nearly all of the laboratories were out of media, and their work almost came to a halt. I then suggested to Burnazyan that we might manufacture nutrient media on the spot. There was nothing else to do so my suggestion was adopted. For media brewing we chose one of the dining rooms, and it took us literally two or three days to resupply ourselves. The problem was solved and there was no further breakdown of supply.

The media brew house was very conveniently situated and it attracted many chefs, owing not only to the media, but also to excellent meat broth (we were sent first-class meat) and alcohol!

Our experience in organizing the production of nutrient media was adopted for the "inventory" by the staff of the national civil defense body, and even made into a film! Indeed, the problem of supplying the country with good-quality nutrient media is still unsolved. This matter has been referred several times to the very highest level. There were even decrees adopted for the construction of special factories, but as soon as an emergency was concluded, the entire issue was dropped.[5]

On my way back to Rostov, an incident occurred which, had things turned out differently, could have had a profound effect upon my future. While waiting for the aircraft in Tashkent I was approached by an unknown young man who handed me an order from Burnazyan[6] to fly immediately to Moscow. Much as I wanted to get back home I had to comply with this order. Upon my arrival at our meeting, Burnazyan stunned me with the proposal that I should take up the post of chief medical officer of health—a deputy to the minister of health, which B. V. Petrovskij had recently become. After speaking to the minister, I was allowed to go home with the warning to keep our conversation secret. I was soon summoned back to Moscow to face the Central Committee of the Communist Party, without whose agreement no one could become the director of a large institute. It turned out that the matter had already been decided, and my wife and I began quietly to get ready for a fresh move.

But as they say, "Man proposes, but God disposes." One day toward the end of 1965 I was reading the *Medical Gazette* when I came across a report that Gen. P. N. Burgasov had been appointed as Chief Medical Officer of Health. Nobody thought of apologizing to me. I learned afterward that this was the normal practice when making appointments: talks were held with several candidates at once. I took this turn of events calmly—I had never thought of being named to the post anyway—and had never been fully convinced of my promotion. Still, the episode left an unpleasant taste in my mouth. This was the first time I had, however inadvertently, crossed swords with General Burgasov, and it was not to be the last. Burgasov was certainly suited for the post; he was older than me, and more experienced. He had previously occupied a number of senior posts.

That would have seemed to be the end of it, but no. Burgasov, having learned that I had been a candidate, regarded me as someone trying to stand in his way. He therefore proceeded to exact his revenge. The first thing he did was to refuse to confirm the institute's work program for the next year on the pretext that it involved "too much biochemistry." He demanded a reorganization of the institute. He then raised a fuss about an internal laboratory case of one lab assistant who had become infected with anthrax. I was not to blame for this incident, although Burgasov demanded my dismissal. I got through it with a severe reprimand, which he later refused to lift. (Under the Labor Code of the USSR these penalties have to be lifted after one year.) After that Burgasov also refused to confirm the appointment of S. Rassudov,[7] my new deputy, in whose lab the infected assistant was working. While I was in a sanatorium in the Crimea recovering from tuberculosis, he appointed M. T. Titenko, retired from the army, to replace Rassudov. Fortunately Titenko turned out to be a decent man. He came to introduce himself to me in the Crimea and to apologize for Burgasov's action.

My relations with Burgasov continued to be unpleasant for many years. There may have been many reasons for his enmity toward me, but he never explained them. For my part, I had no love for Burgasov either, in particular because of his adherence to Stalin and the notorious Lavrenty Beria, the head of the KGB and Stalin's close associate, who was executed soon after Stalin's death. Besides, I have never considered Burgasov a scientist.

Incidentally, Burgasov became well known in the West because of his role in the 1979 anthrax "outbreak"—actually a bioweapons accident—in Sverdlovsk, now Ekaterinburg, in the Ural Mountains. As is now well known,[8] the epidemic resulted from damage to the aerosol equipment in the Sverdlovsk Military Institute, a secret bioweapons laboratory that has to this day never been open to the West. A plume of aerosolized anthrax was released into the air, apparently because a technician had forgotten to change a filter. Sixty-eight civilians died from inhaling anthrax as a result of that accident. At that time, Burgasov was the chief medical officer of health for the Soviet Union. Before his appointment at the Ministry of Health, Burgasov had worked in the Sverdlovsk Military Institute.[9] He therefore was anxious to help his former colleagues to hide their guilt. In this, naturally, he had the clear direction of the government and the approval of Boris Yeltsin, future president of Russia, who was then the First Secretary of the Sverdlovsk Oblast Party Committee.

They thought out an "explanation": the Sverdlosk civilians had died by eating tainted meat! (They ignored the obvious fact that to display the symptoms of inhalational anthrax,* they would have actually had to inhale their food!) But General Burgasov maintained the tainted meat explanation both within the Soviet Union for public consumption, as it were, and abroad to Western scientists. It is important to point out that even today, well into his eighties, General Burgasov still insists on this explanation![10]

Eventually, for a time, Burgasov became more amicable toward me. Apparently he either had forgotten his past deeds or he realized that the competition between us hadn't been my fault. His improved disposition toward me coincided with his appointment as editor of the journal *Molecular Genetics, Microbiology, and Virology*. About these subjects he hadn't the slightest idea. This is another example of the paradox of those times. The founder of this journal had been the USSR minister of health, while Burgasov—an epidemiologist by profession and experience—was still one of the deputy ministers. His appointment was absurd, of course,

---

*As opposed to cutaneous anthrax (contracted through the skin) or intestinal anthrax (contracted through eating infected meat).

but such placements in the USSR were not uncommon. Burgasov's appointment was only formal; as everyone knew, the true experts in the problems were on the editorial board. However, to Burgasov's credit, he always behaved very modestly at meetings of the editorial board and never tried to apply any pressure.

But life went on and everything went along its course. I simply disregarded Burgasov's order for the reorganization of the institute. Actually, I extended the scope of our work in biochemistry and genetics, mainly to do with vibrios, which rapidly bore fruit. By some bureaucratic fluke, the institute contentedly passed one or two years without any confirmation of its program. Thus, in the interstices of power I was able to quietly work.

I had never cared for paperwork, and, once, years earlier while I was still in Saratov, I found myself in an awkward situation over this. Each year Expert Commissions at the Mikrob Institute in Saratov—composed of directors of the antiplague institutes, their deputies, and the heads of the plague-control stations—considered the programs and reports. At one of the plenary sessions I myself, rather than my deputy (at that time M. S. Drozhevkina, an intelligent but very sharp-tongued woman), was called on to present the report. Although hardly anyone was listening to me, my report received all-around approval. When I asked her whether she liked my presentation, Drozhevkina replied: "It all went very well, except that you were reporting last year's figures!"

In Rostov, I devoted particular attention to nutrient media. Soon we had succeeded in developing several original media, producing an actual counterpart to one of the foreign kinds (which we did not have funds to buy). We also managed to prepare a large quantity of dry media for stock. As we have seen, in the case of cholera, for example, large quantities of nutrient media are needed for diagnosis. In the USSR there was a permanent shortage,[11] which the Rostov institute tried to fill. But one sort of nutrient media is not sufficient. Bacteria of different species have different nutritional requirements. One cannot therefore use a single medium for all, so we needed to test and develop many different formulas. Dry media have the advantage of easier storage and are also less likely to spoil; they can be reconstituted by adding water.

I also developed a considerable interest in issues of bacterial systematics or taxonomy. The result of this interest was a new book,[12] which I wrote mostly while convalescing in various tuberculosis sanatoriums. My interest in taxonomy had arisen for two reasons: the need to improve the differentiation between cholera and noncholera vibrios,[13] and in 1966 I had become a member of the international subcommittee on the taxonomy of pasteurellae, which were some of the species of bacteria with which the plague microbe was at that time thought to be associated.[14] Prof. Werner Knapp of the German Democratic Republic (East Germany) was appointed chairman of this subcommittee and Prof. Henry Mollaret of France was named its secretary. Thereafter I used to meet them often and I have maintained friendly relations with W. Knapp ever since (he died several years ago).*

In 1970–71 a cholera epidemic broke out with renewed force, covering almost the whole of southern Russia and the Ukraine. Its biggest foci were in Astrakhan on the Volga River and in Odessa on the Black Sea. It even reached Rostov and Taganrog, though there it was confirmed in only a few cases, and probably it was brought in from Astrakhan or Odessa. By insisting on the necessary medical protocols I found myself in a lot of trouble with local authorities. They did not wish to acknowledge that cholera could be present in a city like Rostov, since that might be considered an unsatisfactory state of public health. So they cast doubt on every one of our diagnoses. They were particularly exasperated that I was required to notify Moscow of the outbreak. Rivalry between the USSR Ministry of Health as head of the plague-control establishment, and the Russian Ministry of Health, to which the Rostov Region Health Committee was subordinate, further complicated the situation. It is worth noting that, while Rostov city had a relatively high standard of public health and good conveniences, and therefore was not receptive to cholera, the surrounding Rostov region itself has since become an endemic area for cholera. In the period of 1970 to 1990 more than 303 cases of cholera and 461 vibrio carriers have been registered there, with the great majority coming from Rostov and Taganrog![15]

---

*Mollaret and Knapp were two of the leading plague researchers of the mid-twentieth century.

The staff of the institute was metaphorically torn in pieces. Besides Rostov and its region, they were working in Astrakhan, Odessa, Kerch, and other cities. I myself had to work in Makhachkala and Donetsk, not to mention periodic journeys to Mariupol, Kherson (both in the Ukraine), and Kerch (Crimea). Over three thousand cases of cholera and vibrio-carriers were recorded. For its scientific and practical achievements in combating cholera the Rostov Institute during my time in 1971 became the "leading" institute for this problem in the USSR. It has essentially remained so up to now, as confirmed in particular by the publication of the fundamental work on the spread of El Tor vibrions in bodies of surface water and effluents in the USSR during the seventh cholera pandemic[16] and by its active share in eliminating the major cholera epidemic in Daghestan in 1994.

This last epidemic in Daghestan should come as no surprise. Over a quarter of a century since the first recorded outbreak there, very little had changed in the public-health situation, including in the capital, Makhachkala. Daghestan is a remote area in the Caucasus Mountains, and the public-health system there is very poor, especially in the country. Most houses do not have either sewage systems or a centralized water supply. Exacerbating the situation, many Dagestanis are devout Muslims, and many take part in mass pilgrimages to Saudi Arabia. All of these factors enabled cholera to take permanent hold after the epidemic in south Russia of 1970–1971 allowed the disease to enter the region. But for the authorities it is simpler to declare that cholera is brought in from outside Russia than to build water and sewage systems: there is no money for that.

I still have the photographs of certain "vulnerable areas" of Makhachkala, which show that there was no need to blame outside sources for the epidemic, as Russian authorities have recently done in attributing the outbreak to pilgrims from Saudi Arabia.

Apart from my duties in scientific administration, much of my time in Rostov was taken up with public activities as a deputy, first of the district, and then of the Region Council of Workers' Deputies, which at that time included the writers M. Sholokhov and V. Zakrutkin, and also Yu Zhdanov, the son of A. A. Zhdanov, and one of the husbands of Svetlana

Allilueva (Stalin's daughter). I was elected a deputy in a general election, in which the population of Rostov oblast took part. But the Provincial Party Committee put me up for election. As a director of the institute and a Party member I had no choice, I had to consent. All elections to the Party and to Soviet organs were then uncontested ones: only one candidate was ever on the slate. Also I must say that to be a deputy was honorable, but absolutely useless, since as a rule nothing depended on deputies. All decisions had been made ahead of time. It was a waste of time. Since my Irkutsk period, my attitude toward councils of various levels has not changed; I regard service as a deputy a pointless chore, especially since no one ever takes any notice of the opinions of deputies. We served as a screen behind which the authorities acted as it suited them.

The sessions were tedious and dull, although it was "strictly mandatory" to turn up. The doors were usually locked after the start (and until the end). Nevertheless I managed to get away. To do so I usually took off my overcoat in the car and asked my driver to drive up at the break time. After telling the "boys" who guarded the entrance that I was "going out for a smoke," I would leave and get into the car. I used the same tactic at the numerous sessions of the "Party Economic Activist Groups" which were attended by the managers of enterprises of every size of the city and the region. It was curious that during the actual sessions one could read, chat, or even sit with friends in the buffet!

Much more agreeable were the meetings of the Rostov Region section of the All-Union Microbiological Society, which I organized. The Microbiological Society allowed us to communicate with colleagues from other institutes, and to discuss problems of mutual concern and new scientific methods. The society had a right to publish as well.

A good part of my effort also went into construction: in one instance, I was involved in the building of a large housing block in the center of Rostov for members of the institute. Anyone who has ever dealt with such matters will realize how difficult they are.

My years living in Rostov after the events in Nukus were closely linked to a genuine Cossack, a marvelous woman and a great scientist: Zinaida V. Ermoleva, who loved Rostov passionately and cared for its inhabitants. Ermoleva was herself an interesting and erudite person. Born in Rostov, she pioneered the use of penicillin on the battlefield (1941–45) shortly after its discovery by the Scottish Dr. Alexander Fleming (the Nobel Prize winner). She was also a great expert in cholera. In the twenties she proved that a luminous vibrio found in water is a form of Vibrio cholerae. The heroine of a novel titled *The Opened Book* (Otkrytaya kniga), by V. A. Kaverin, was based on her.

Ermoleva had a particular fondness for the Rostov Antiplague Institute and, it seems, for me as a scientist; in particular Ermoleva seconded my candidacy during election to the USSR Academy of Medical Sciences (AMS of the USSR), for which I am indebted to her.[17] My wife and I are proud of being counted among her friends.[18]

As in Irkutsk, our home in Rostov was open to our friends and acquaintances, and we had plenty of visitors for nine years. In spite of all my work, I had enough time for sport (swimming, rowing, table tennis, and badminton). Looking back, I now realize that this was the brightest time in my life. I also have affection for Rostov because my family's original home, Taganrog, was nearby.[19] We traveled several times with my mother to Taganrog, Mariupol, and Novocherkassk trying to find the village near Matveev Kurgan[20] (not far from Rostov) where she had spent her youth. My wife and I toured all over the Caucasus, and reached Crimea by ferry, and we usually took our children to the Black Sea in the summer.

How things would have turned out later in Rostov, and whether we should have willingly left it all behind, one can only guess. I have always felt that one cannot stay for long in one place. I think that for any success, and for one's study of science, too, from time to time it is necessary to have a change of atmosphere, new impressions, and, yes, a new job. I may have been right, though the notion suffers from one weakness, which may be the most important consideration for a scientist: a change of place is inevitably linked with a change of course in one's work, but outstanding success can be achieved only by those who keep to the same line.

But the same line—the peaceful pursuit of science, the efforts to control plague, cholera, and other epidemic diseases—is just what I was about to give up. The work I had done on the cholera epidemic and my administrative successes in Irkutsk and Rostov had brought me into the purview of people in the Soviet Union who had quite different interests. Gradually I would find the course of my life being directed down an entirely different stream, one I had initially never thought to follow.

NOTES

1. The Directorate controlled a network of special hospitals and medical units to serve biological weapons research-and-development facilities. A second network investigated biological agents that could cause nonlethal and lethal organic and physiological changes (Program "Flute"). Several labs in this second network developed toxins and other substances for use against "individual human targets."

2. It was clear that cholera El-Tor was the dominant strain in Nukus.

3. What the true source of the epidemic was, it is difficult to say exactly even now. But if it was Afghanistan, why did the outbreak mainly affect Nukus, which is located quite far from the Afghanistan border? On the other hand, just then vibrios were isolated from water reservoirs in many regions of Uzbekistan, which is close to Afghanistan, as well as other republics of the USSR.

4. SEATO, the Southeast Asia Treaty Organization, a military and political alliance established in 1958 and dissolved in 1977. The alliance was made up of Australia, France, Great Britain, New Zealand, the Philippines, Thailand, Pakistan (through 1973), and the United States. Analogous to NATO, SEATO members pledged to defend one another against military aggression in Asia. It was formed in large measure to prevent the expansion of Communist influence in Southeast Asia. Therefore, the Soviet Union viewed SEATO unfavorably. (Ben Garrett)

5. Incidentally, even Glavmikrobioprom has failed to sort the matter out, although it possessed every facility, as I describe below.

6. Apparently my work in Nukus had favorably impressed him; I also was beginning to develop a reputation as a scientist and as an administrator.

7. See page 100.

8. See M. Meselson, "The Sverdlovsk Anthrax Outbreak of 1979," *Science* 266, no. 5188: 1292–98.

9. See his article "Smallpox Is Better Weapon than Anthrax" (in Russian), *Moscow News*, November 13, 2001.

10. His enmity to me, which had waned for a while, become especially strong again after the publication of the original Russian version of this book, *Perevertish*, in 1995, as well as some other journalistic publications.

11. See note 3.

12. I. V. Domaradskij, *The Agents of Pasteurellosis and Related Diseases* (in Russian) (Moscow: Meditsina, 1971).

13. Sometimes differentiating between cholera and noncholera vibrios is very difficult. For improvement of diagnostics it is necessary to learn microbes in detail. It needs a deep study of all the characteristics of microbes.

14. At that time there was a heated dispute about the relationship between plague microbes and some related bacteria (Pasteureulla). By the way, plague microbe was considered then to be a member of the genus Pasteurella (now it is classified as a member of genus *Yersinia* from another family—Enterobacteriaceae). My work was devoted to a comparative study of the plague microbe and related bacteria.

15. "Cholera" (proceedings of a Russian scientific conference, Rostov-on-Don, November 18–19, 1992)).

16. G. M. Medinskiy, M. I. Narkevich, Yu. M. Lomov, et al. *Spravochnik-kadastr rasprostraneniya vibrionov eltor v poverkhnostnych vodoemakh i stochnykh vodakh na territorii Sovetskogo Soyuza v 7-yu pandemiyu kholery* (A reference survey of the spread of El Tor vibrions in bodies of surface water and effluents in the USSR during the 7th cholera pandemic), ed. Yu. M. Lomov (Rostov-on-Don: Antiplague Institute Publishing House, 1991).

17. See Appendix 1: "Academies."

18. Among them were V. D. Timakov, G. P. Rudnev, I. A. Kassirskij, P. N. Kashkin, and several other remarkable personalities. (See Glossary of Names.)

19. See chapter 1.

20. "Kurgan" means "barrow" or "burial mound"—apparently the Matveev Kurgan was the site of a burial mound of the long-vanished Kurgan people, an Indo-European-speaking group who once inhabited much of Russia and the Ukraine; by 2000 B.C.E. they had spread across Europe into Germany.

CHAPTER 6

"PROBLEM NO. 5"

He who says A must also say B.

H. Heine

A s I said in the previous chapter, my move to Rostov coincided with the transformation of that institute, which at that point began to emerge as the leading institute in the plague-control system on "Problem No. 5." My own deepening involvement—the gateway for me into the world of biological weapons—began here at Rostov with Problem No. 5. This odd-sounding code word stands for a Soviet program for the "antibacterial protection of the population," or, in other words, the development of means of protection from biological weapons. Almost all civil microbiological institutes of the USSR were involved in Problem No. 5. Since the early seventies, when the Biological and Toxin Weapons Convention (BTWC) was signed by the Soviet Union, the United States, and (as of January 1, 2001) 144 other nations,[1] the emphasis of Problem No. 5 shifted. It began to serve as a "legend," or cover story, to hide the illegal and ultrasecret biological weapons program known as Problem "Ferment" (or Problem "F"). With the signing of the BTWC, biological defense was no longer a priority; under cover of the treaty, and unknown to the rest of the signatories, a greatly expanded and ambitious bioweapons program would rapidly take root in the USSR.

Developing such a program became the main task of a supersecret group, the Interagency Scientific and Technical Council for Molecular Biology and Genetics. Problem No. 5 became little more than a smoke-screen for these hidden and dangerous activities. It was used both by the Ministry of Defense, and by a new organization, Biopreparat, the "civilian arm" of the biological weapons program.

But when I first encountered Problem No. 5 in the early 1950s, while I was working at the Mikrob Institute, its program was still defense-related. This was during the Cold War when, as we well knew, the United States had an active biological weapons program. Biological defense was therefore imperative, and I was given the task of developing a fast method of detecting the plague microbe. The main method of detection of plague as well as other bacteria is to isolate them on different liquid or solid nutrient media. There are also some other methods, such as the use of specific antibodies to see whether the bacteria bind to them. One of these methods was developed in Rostov, as we saw in the preceding chapter. But none of these methods can detect bacteria rapidly. I worked out a specific indirect method to test for the presence of plague microbes, which only took four or five hours. The method was based on the following principle: when a phage (a bacteria-killing virus) particle meets a bacterial cell, it enters the cell and multiplies there. Thus, during multiplication of the phage, the phage's titer (or concentration) increases. The growth of a phage titer thus reveals the presence of plague microbes: the more phage detected, the more plague microbes there must be.

This task appeared to be very prestigious (after all, it was defense-related), and it was with pleasure that I set about tackling it. Before I hit on the idea of phage titers I tried everything under the sun to develop a method of rapid diagnosis! I even tried using isotopes, which were only just becoming available. Isotopes are radionucleids, employed widely in biology as indicators. The Institute of Advanced Medical Studies taught different new methods suitable for practical medicine, including radiology, making use of isotopes. We thought isotopes were amazing then. The Mikrob Institute sent me to Moscow to attend courses on isotopes at the Institute of Advanced Medical Studies. I stayed for nearly four months (and, incidentally, had the time to have my doctoral thesis approved by A. E. Braunshtejn, whom I have mentioned previously.)[2]

But I failed to get any positive results by using isotopes as a form of rapid diagnosis. It finally occurred to me to try using the growth of a phage titer. This looked like a fruitful idea, but a new problem arose: it was very difficult to determine the increase of a titer in a liquid medium. I tried liquid medium since at that time it was a traditional method, but I had to reject the method since I could not obtain satisfactory results. I needed to learn how to do phage titration in a solid medium; there would have been no opportunity to learn such a technique in Saratov, where no one knew the technique, and we had no access to foreign literature.

But it so happened that Professor Fenyuk of the Rostov Institute had gotten to know D. M. Goldfarb in Sochi[3] whom he had told of my research. Goldfarb was likewise engaged on phages, and via Fenyuk he invited me to Timakov's[4] laboratory where he was then working. I look back with pleasure on my stay at Timakov's laboratory in Furkasovskij Lane. I came to know a great many interesting people and I became skilled at phage titration using the Grazi (two-layer) method. Thanks to that, after my return to Saratov the method was ready. I remember the joy my colleagues and I felt when for the first time we could see, after only two or three hours, negative colonies of plague phage on our dishes—the bacteria-eating viruses had destroyed the bacterial colonies! This meant that, using phages and the two-layer phage titration method I had learned at Timakov's laboratory, I could determine whether any particular bacterial colonies were actually plague colonies by seeing whether phages could destroy them.

A special instruction was written on the basis of this method, although it was stamped "Secret." At that time much research devoted to plague and its causative agent *Yersinia pestis,* as well as some work on Problem No. 5, was classified; among these restricted works was my research on rapid diagnostics of plague. However, being young I attached no importance to such designations, and once I had become the director of the Plague-Control Institute in Irkutsk, where censorship was not so severe, I issued this instruction in a separate brochure. I managed to get away with it then and had no further professional trouble, although this method later became current among all the people responsible for work on the rapid diagnosis of dangerous bacteria. Unfortunately, this issue strained relations between Goldfarb and me: he felt that credit for the method should have been attributed to him. I published this method before he did, and I have documentary evidence from the Mikrob Institute to prove it. But relations between Goldfarb and me did not improve. It may also be that Goldfarb had some right on his side, since had I not been posted to work alongside him I might never have achieved my results.

Goldfarb was known for having a somewhat ridiculous character and an exaggerated sense of his own importance. Having fallen out with V. Timakov, who had treated him very well, and having twice failed to be elected to the USSR Academy of Medical Science, Goldfarb, along with his son, joined the ranks of the "dissidents." Often he quarreled with people for no reason whatever. He publicly jeered at colleagues and even at his friends. Of course, that was reflected in the results of the Academy elections. When his son emigrated to the United States, Goldfarb visited him a few times but for a long time never stayed. It may be that in spite of his battles with the Soviet authorities he remembered the war, in which he had lost a leg, and he still retained some trace of patriotism.

Much later, Goldfarb was involved in a peculiar matter that, I am glad to say, involved me only indirectly. In the spring of 1984, while his luggage was being examined at Sheremetevo Airport, a packet of cigarettes that looked suspicious to the customs people was taken from him. I don't know why. It may have had something to do with Goldfarb's job at the Institute of Genetics of the Soviet Academy of Science. These cigarettes were sent to the All-Union Institute of Applied Microbiology with the request for a bacteriological analysis. As deputy director of this institute, I handed this job to my colleagues, and from the cigarettes they isolated some genetically altered cultures of an intestinal bacterium, a frequent subject of genetic research.

For a long time Goldfarb had been engaged in the study of bacterial genetics; he had advanced quite far in this field. Naturally, when he decided to move abroad he did not want to lose some of the genetically modified strains he had created, but removal abroad of such strains was prohibited. Apparently these cultures were of some scientific interest, and merely for that reason they could not be sent away without a special permit. Therefore he thought up this unusual way of transporting his strains. Whether they had any scientific interest, I cannot say.

Goldfarb was deemed to have thought up this original means of smuggling. However, as I learned later, he had categorically denied the charge, declaring that, since he knew nothing about these cultures he could not have acceded to the request from "a certain woman friend" to bring these cigarettes into the United States. One way or another, the story was spread about and was mentioned on the *Voice of America* (May 11–12, 1984) in the usual "Cold War" fashion for those days. But certainly Goldfarb was not a spy![5]

Of the many investigations into Problem No. 5 on which I worked during the Rostov period, I consider the best to have been the production of an EV strain of plague which was resistant to the most widely used antibiotics; its use as a basis for a dry vaccine[6] was subsequently granted "wide recognition." The EV strain of *Yersinia pestis* was isolated in Madagascar nearly seventy years ago by French scientists J. Robic and G. Girard. At first it was virulent. Subsequently it lost its virulence, but retained immunogenicity—in other words, its ability to protect against virulent plague infection. Using EV we developed an antibiotic-resistant plague strain that could be used for vaccination simultaneously with the administration of antibiotics; under normal circumstances, of course, antibiotics would kill EV vaccine as well as virulent plague. I am not aware of the existence of any such vaccine abroad.

The development of an all-round resistant EV strain was made possible because we were then seriously tackling the genetics of the plague microbe. Some of this research, together with the biochemical results, formed the basis for a book that I wrote in conjunction with my colleagues E. P. Golubinskij, S. A. Lebedeva, and Yu. G. Suchkov.[7]

We developed the polyresistant vaccine for civil defense purposes. We considered that the polyresistant strain might be useful in case of a need for immunization in combination with antibiotics (a kind of "urgent prophylaxis"). In Russia, we consider such live vaccine strains of *Yersinia pestis, Francisella tularensis,* and *Bacillus anthracis* as the essential means of defense against biological weapons. Vaccines provide safety against a disease for much longer than antibiotics. If there is a renewed threat, exposed people must take whole courses of antibiotics again, while a vaccine protects at least for an entire year and often for much longer.

On the other hand, there was a dark side to that research, though I did not realize it at the time; Problem No. 5 was still concerned with biological *defense,* and the Rostov Institute (as well as all other plague-control centers) were never directly involved in the development or the production of biological weapons. But antibiotic-resistant *virulent* bacterial strains, developed using the same technology as we used for our vaccine strains, were considered by our military as a means of attack and a powerful weapon.

Beginning in Rostov, my colleagues and I began to make other significant discoveries on plague genetics, which could not be published because they were deemed secret. The inability to publish was a major personal and scientific loss for me, though it was the natural and unfortunate consequence of my engagement in secret research. One of my colleagues, Elena Koltsova, demonstrated the first indication of the presence of plasmids in the plague microbe (1970):[8] she succeeded in transferring the genes that produce pesticin—one of plague's virulence factors that is found on a plasmid—from the plague microbe to intestinal bacteria. Despite her discovery, American plague experts R. Little and R. Brubaker (1972) claimed at the time that the plague microbe had no plasmids—a claim they eventually recanted. When I moved to Moscow, I continued research along the same lines with colleagues at the Kirov Military Institute. As a result of our joint efforts, plague plasmids were detected. What is more, it was possible to establish a link in pathogenicity (virulence) of the plague microbe with these plasmids. Since then, the facts of the existence of plague plasmids and their connection to virulence have become accepted scientific fact. It is important to note that abroad the first confirmation of this began to appear nearly a decade later in 1980–1981! But the data attesting to our prior research was top secret, and therefore we were not recognized as having made the discovery.

My research at Rostov under the aegis of Problem No. 5 produced one other important result: the creation of SPECTs—specialized epidemic control teams—which I feel have played, and continue to play, a major part in combating highly dangerous infections. The conceptual inspiration and a direct participant in organizing these SPECTs was my friend Prof. G. M. Medinskij, while I carried out the design and the administrative work to create them. The SPECTs themselves, as originally planned by the Soviet Ministry of Health (in 1964), were mobile nonmilitary civil defense formations mainly intended for wartime, which originally confined their range of duties to the specific indication of biological weapons. These SPECTS were set up like little institutes that could be deployed very quickly wherever they were needed. They had everything required for microbiological diagnostics and they were suitable for peacetime outbreaks as well as during wartime.

For my part in the cholera epidemic control measures I was awarded the Mark of Honor and the Order of Lenin, the highest award in the USSR, and I was also granted the title of "Honored worker (of science and technology of the Karakalpak ASSR)."

The work on Problem No. 5, together with my efforts in the war on cholera—first in Nukus in 1965 and later in Makhachkala, Donetsk, and other cities (1970–71)—led to my transfer to Moscow to work at Glavmikrobioprom, the Main Directorate of the Microbiological Industry under the Council of Ministers of the USSR. This directorate is so important that it warrants a brief note of explanation, or subsequent developments to be discussed may be difficult for the reader to follow.

V. D. Belyaev created Glavmikrobioprom initially for the production of agricultural foodstuffs (mainly proteins of yeast grown on paraffins and gases, which were fed widely to cattle and poultry) and some medicines (xylitol—a nonsugar sweetener intended for diabetics—insulin, antibiotics, enzymes, and so on). Belayev, a very clever man, was a chemical industry researcher involved in the creation of a production system for a chemical weapons agent, a poison gas called Sarin. I believe he always was connected with the Ministry of Defense and the Military Industrial Commission. He created Glavmikrobioprom in particular as a cover for connections with the Ministry of Defense and the Military Industrial Commission, which I was to learn about when I moved to Moscow. In 1973, to provide civilian cover for advanced military research into biological weapons, another directorate, this one under Glavmikrobioprom, was created called *Biopreparat.*[9] This directly followed the signing of the Biological and Toxin Weapons Convention in the same year. The entire Soviet biological weapons program went underground, and some kind of cover was needed to hide the fact that illicit research was going on. Biopreparat provided the bureaucratic cover. The majority of Biopreparat's personnel initially came from the army's Fifteenth Directorate, which kept this supposedly civilian directorate effectively under military control. A government reorganization in the mid-1980s transferred Biopreparat to the Ministry of Medical and Microbiological Industries, but it continued to enjoy virtually autonomous authority as the principal government agency for biological weapons research and development. Biopreparat was officially responsible for civilian facilities around the country dedicated to the research and development of drugs, vaccines, biopesticides, and some laboratory and hospital equipment, but many of its facilities also served as biological weapons development and production plants, and were earmarked as "reserve" or mobilization units for use in case of war. It is very difficult even for on-site foreign inspectors, but especially from space via satellites, to distinguish facilities that produce only vaccines from those making offensive biological weapons, so the cover Biopreparat facilities provided was plausible.

The first head of Biopreparat was Gen. V. I. Ogarkov; but from 1979 until only very recently Gen. Yuri T. Kalinin headed it. Ogarkov was also a deputy chief of all of Glavmikrobioprom, as well as the head of the supersecret Interagency Science and Technology Council on Molecular Biology and Genetics, about which I will have more to say later (see chapter 8).

In the winter of 1972, while I was resting in a sanatorium outside Moscow recovering from tuberculosis, a car unexpectedly arrived to take me to the Ministry of Health USSR. I had been summoned to see A. I.

Burnazyan, who announced that the powers that be had decided to transfer me to Moscow. After speaking to someone on the phone, he took me to the Kremlin and introduced me to D. P. Novikov, the head of a department of the Military Industrial Commission (MIC). It was from Novikov that I first learned about the creation of the Interagency Scientific and Technology Council on Molecular Biology and Genetics, the new organization at Glavmikrobioprom under the USSR Council of Ministers. He told me that they needed someone with my working experience, which would be guaranteed by a recommendation from Burnazyan, the deputy Minister of Health, whom I had known from the epidemic in Nukus. (Apparently they were impressed with my work in Rostov and Irkutsk and with my knowledge of cholera epidemics.) I had been introduced by the Ministry of Health to the Directorate of Microbiological Industry as an expert in the fields of general microbiology and biochemistry and perhaps, also, as an expert in medical microbiology. V. D. Belyaev, the head of Glavmikrobioprom, also took part in this conversation. I do not know in any more detail what, exactly, they intended for me. But certainly at that point I was not summoned to work directly on biological weapons. Biopreparat did not arise for some time after my appearance in Moscow, and the Glavmikrobioprom had neither specialists nor the necessary conditions to work on biological weapons. My selection was carried out in an atmosphere of top secrecy, but this was true for any leading position in the USSR.

Later, as I was walking with Belyaev through the Spassklj Gate on to Red Square, he asked me whether I knew how to write. I couldn't conceal my surprise, and began to list all my publications and books. Belyaev laughed but said nothing. I realized the meaning of his question, and of his laughter, later, when I had to submit to him the relevant papers for signature. As a rule none of them would ever satisfy him, either for style or content. It turned out that I really don't know how to write, at least not the type of official documents that Belyaev had in mind. Furthermore Belyaev, for one reason or another, could not always plainly say what he meant.[10] I can remember going to Belyaev many times with the same draft letter to the Soviet Ministry of Defense, but I simply never managed to please him. Belyaev died in 1979, and the letter remained unsent. I learned later that the reason for this lay hidden under political considerations, which I was in no position—then or now—to understand.

Negotiations for my transfer to Moscow dragged on for almost a year. I refused to endure again the difficult conditions I had faced when moving to Rostov; I therefore set out a number of conditions concerning a flat in Moscow, the nature of the work, and a job for my wife, to which I invariably received an answer directly from Belyaev (an unprecedented case in view of Belyaev's high position in the hierarchy; I still have his letters). I should add that Belyaev did all that he had promised, except for finding work for my wife.

In 1972, in order to justify my new appointment, even though I was already a corresponding member of the USSR Academy of Medical Science, I defended a second doctoral thesis, this time as a candidate for a doctorate of biological science. I thought then that another Ph.D. might be more suitable for work in such an organization as the Glavmikrobioprom. This met with neither comprehension nor approval in academic circles ("What's the idea? One degree only is enough for us to be who we are"). Strange to say, this second doctoral defense was also held at the Mikrob Institute exactly sixteen years after my first defense, and on the same date, June 16.

After moving to Moscow I had only indirect contacts with Problem No. 5; and it receded into the background, serving, as we will see next, as a cover for a new operation, one I did not yet know about when I moved to Moscow. Had I known what was in store for me I would not have wanted to do it, and I would certainly have refused.

NOTES

1. See M. Leitenberg, "Biological Weapons: Scientist and Citizen," special issue, *Chemical and Biological Warfare* 9, no. 7 (1967): 153–67.

2. This approval was very important to me professionally: Braunshtejn was a world-renowned biochemist and his positive evaluation of it was very important for my doctoral defense.

3. A spa on the shore of the Black Sea, in Abkhazia.

4. See Glossary of Names.

5. The Goldfarb Affair has been understood differently in the West. The son, microbiologist Alexander Goldfarb, was permitted to emigrate in 1975, winding up in the United States as a distinguished professor at Columbia University. In 1979 the father, then retired, was granted permission to emigrate. But when he attempted to leave, he was informed that he could not depart the Soviet Union because he possessed classified information—in the form of his memories and recollections of his research. After being denied permission to emigrate, David Goldfarb became a "refusnik"—a term applied to such Soviet luminaries as Nobel Prize-winning physicist Andrey Sakharov, noted for their refusal to go along with the Soviet system. In 1984 Goldfarb was finally granted a permission to emigrate. When he arrived at Moscow's Sheremetyevo airport, guards seized a pack of cigarettes, claiming that they looked suspicious. Goldfarb claimed the cigarettes were a going-away present from a female friend. The Soviets later reported that analysis of the cigarettes showed some genetically modified bacteria had been secreted within the cigarettes. The truth of this situation is unlikely to be known with certainty, but the record of abuses from the Soviet KGB and other intelligence agencies suggests Goldfarb's claim that the cigarettes were truly a gift is legitimate, and the lack of transparency surrounding the Soviet laboratory's analysis of the cigarettes casts doubt on their purported results. Goldfarb's situation continued to be a matter of international concern, resulting in high-level appeals to the Soviet government to permit him to depart. These appeals worked. In October 1986 famed industrialist and philanthropist Armand Hammer dispatched his private jet to Moscow, and David Goldfarb was permitted to depart. He settled in the United States. (Ben Garrett)

6. At present almost all vaccines are dried under vacuum pressure at low temperatures. Under these conditions microbes survive. Dry vaccines are kept for years; liquid vaccines can spoil.

7. I. V. Domaradskij et al., *The Biochemistry and Genetics of the Plague Microbe* (Moscow: Meditsina, 1974).

8. Plasmids are loops of extrachromosomal DNA that were initially discovered by American scientist Joshua Lederberg—for which he was awarded the Nobel Prize in 1951 at the age of thirty-one. In plague, these circles of DNA contain most of the bacterium's virulence factors—the factors that make the microbe so deadly.

9. Biopreparat, like many other secret Soviet organizations, was also known by a so-called Post Office Box number—in this case P.O. Box A-1063.

10. I had to prepare for Belyaev letters to different institutions (the Central Committee of the Communist Party, Council of Ministers, Military Industrial Commission, and others). Writing official letters like these demands a particular official style that is different from, for example, a scientific one. I lacked the requisite experience for this bureaucratic task. Besides, Belyaev often could not exactly express his thought or general sense. Neither at the Ministry of Defense nor at Glavmikrobioprom did the chiefs write anything for themselves.

CHAPTER 7

A SHORT HISTORY OF BIOWEAPONS IN THE SOVIET UNION

This is an old story . . .

H. Heine

I t is difficult to say who first conceived the idea of deliberately infecting people and cattle, but ancient books and chronicles are full of accounts. For example, we read in the Old Testament, "And I will bring a sword upon you that shall avenge the quarrel of my covenant. . . . I will send the pestilence among you; and you shall be delivered into the hand of the enemy" (Lev. 26:25) or "For I will punish them that dwell in the land of Egypt, as I have punished Jerusalem, by the sword, by the famine, and by the pestilence" (Jer. 44:13).[1]

Repeated cases are known in history of the poisoning of the enemy's wells, which would now be considered as instances of biological warfare. However, those are more or less exciting legends, and often mere fiction, like *Alpugara*, a Polish novel by A. Mitskevitch, which describes the perfect model of biological sabotage in the very core of the enemy's camp. There is also the commonly told story of how the Black Death first came to Europe in 1348, when, according to Gabriele de Mussi, an Italian chronicler, the Turks besieging the Crimean town of Kaffa, then held by the Genoese, were struck with plague. They then catapulted the bodies of plague victims over the city walls, bringing plague to the city. The Genoese fled by ship and brought the illness with them to Italy and then to the rest of Europe. De Mussi, it is now known, was not an eyewitness to the events he described; no one knows whether the Genoese fell ill because of the plague catapults, or whether infected rats, no respecters of sieges, crossed the city walls instead.

It is only in the seventeenth century that we find the first indisputable use of germs as weapons, the prototypical example of biological warfare. The famous French bacteriologist Charles Nicolle, a Nobel Prize winner (1928),[2] found precise proof of the intentional smallpox contamination of American Indians in the 1763 correspondence between British General Amherst, the governor on New Scotland (North America), and his subordinate, Colonel Bouquet, the Fort Pitt major. General Amherst wrote the following: "Could you try to spread smallpox within Indian rebels? We need to use all means to exterminate those barbarians." Colonel Bouquet answers immediately: "I will try to contaminate them using the blankets, which I will find a way to deliver to them." The British General agrees without hesitation: "I recommend that you should spread smallpox in this way and use all possible means to destroy at last this horrible race."[3]

Nicolle's findings were confirmed by two American authors, E. V. Stern and A. E. Stern, in their book *The Influence of Smallpox on the Fate of the American Indians.** The Sterns found proof that Colonel Bouquet had indeed taken two blankets and a shawl from a smallpox hospital and delivered them to the Indians. Immediately thereafter, a raging smallpox epidemic broke out among the Indian tribes in what is now the state of Ohio.[4]

The first I had ever heard of biological weapons was in reports that the Japanese, who had worked on bacteriological weapons before World War II, had developed and used such weapons during the war. In particular, their aim was the creation of a weapon using *Yersinia pestis,* the bacterium that causes plague. They devised special porcelain bombs filled with fleas (they used the so-called human flea, *Pulex irritans*). As was proved in 1949 during the Khabarowsk trial,[5] the infamous Japanese Unit 731 tested its weapon on people, in particular on captives and prisoners

---

*Boston, 1945.

in Manchuria, near Harbin. Furthermore, it is now known that they also used their plague weapon in China.[6]

During my stay in Ninbo (China, Qejiang province) I was shown a map indicating the areas in which the Japanese in 1940 had dropped bombs loaded with plague microbe. (I managed to take the map with me.) According to Chinese information many people died during these attacks.

But all of the details of the Japanese plague attacks came out much later. My real understanding of the nature of biological weapons came, instead, from a book by Theodor Rosebury titled *Peace or Pestilence.** Rosebury's book explained much that I had not understood from the rumors of the Japanese action. Apparently Rosebury's book also caused Soviet military people to take the problems of biological warfare seriously.

As I gathered from subsequent discussions with specialists, serious work on the development of biological weapons in the Soviet Union began in the early 1940s in what was then Kirov[7] (now Vyatka), at the Research Institute of Epidemiology and Hygiene (RIEH). These specialists were medical officers who were assigned to Biopreparat by the Ministry of Defense. In particular, I spoke with L. Klyucharev and V. Ogarkov, both of whom we shall meet again (see chapter 8). But they did not divulge to me the details of their work on biological weapons at the military institutes.

It seems, however, that despite the impetus the Rosebury book provided, the USSR had actually begun the research and development of biological weapons as far back as the first years of the state. D. K. Zabolotnij, according to numerous accounts one of the first scientific researchers to accept the Soviet regime unconditionally, was also a noted microbiologist and epidemiologist with an international reputation as a plague-control expert. At a conference on cholera called by the Petrograd Soviet of Worker, Soldier, and Peasant Deputies in 1918, Zabolotnij declared that he was placing himself and his expertise "at the disposal of the working class." Zabolotnij submitted a plan for the Kronschtadt plague station[8] on March 9, 1918, at a meeting of physician's colleges, even before the formation of the People's Health Commissariat. His plan at this isolated station was perhaps abetted by the strict compliance with the rules pre-

---

*T. Rosebury, *Peace or Pestilence* (New York: Whittlesey House, 1949).

scribed in the *Codex of Governmental Measures Against the Introduction and Spread of Cholera and Plague within the Empire and over Ground and Sea Borders.** One of the articles of the *Codex* stated that "the conduct of any sort of experiments or research into bubonic plague would be concentrated exclusively on the premises of the Fort of Czar Alexander I at Kronschtadt, and that these studies were prohibited in all other institutions of [St.] Petersburg, in view of the danger they posed."[9] In other words, Russian researchers were free to carry out research into bubonic plague at Kronschtadt.[10]

Zabalotnij became head of the Sanitation and Epidemiology Commission of the Main Military Sanitary Administration (MMSA) and focused the Institute of Experimental Medicine on the control of epidemics, launching mass production of vaccines and serums at that institute. As chairman of the Sanitation and Epidemiology Commission he organized numerous epidemiology units to combat typhus and cholera. But in light of his ideological adherence to the new political regime, I cannot rule out the possibility that D. Zabolotnij, like others later involved in Problem No. 5, may have been involved, at least tangentially, in the initiation of the development of biological weapons in the USSR. During the first years of the Soviet regime, D. Zabolotnij was probably the country's only researcher who knew particularly dangerous infections, including plague, firsthand, so his knowledge would have been invaluable in the early development of an offensive program.

By 1925, the Geneva Protocol banning the use in war of choking, poisonous, or other such gases or biological means, was already in force. The Soviet Union subscribed to the protocol in 1927. But in 1928 a report had appeared abroad that to the north of the Caspian Sea the Soviet Union was deploying a test range for biological weapons, although nobody could confirm it. Whether the report was true or not, evidence of the Soviet Union's possession of biological weapons before the Second World War can be found in K. I. Voroshilov's speech on February 22, 1938: "Ten years ago or more the Soviet Union signed a convention abolishing the use of poison gas and biological weapons.[11] To that we still adhere, but if our enemies

---

*St. Petersburg, 1902.

use such methods against us, I tell you that we are prepared—also fully prepared to use them against aggressors on their own soil."[12]

Certainly by the early 1930s biological weapons were both a topic of interest in the Soviet Union and, increasingly, a subject of research. We had already begun to develop our own biological agents. Nowadays no one can say for sure what those early Soviet biological agents were. But if K. Alibekov[13] is to be believed, they were quite primitive: corpses of infected animals dried and ground to powder. We know that between 1930 and 1940, research into problems related to the preparation for biological warfare was concentrated in the Red Army Bacteriology Institute located in Vlasikha, about thirty kilometers west of Moscow.[14] The institute studied agents of gas-gangrene and tetanus, which are wound infections caused by anaerobic bacteria,[15] botulism[16] and plague, brucellosis and typhus, paratyphoids and others, all of which suggests that these pathogens (germs that cause disease in animals, plants, or humans) were probably the early Soviet biological agents.

The Red Army Bacteriology Institute's director was division doctor I. M. Velikanov. In the fall of 1934 Dr. Velikanov was sent to an international conference of the Red Cross in Japan where he may have held secret talks on biological weapons with the Japanese military. According to his son's account, the next year, under the pen name of Ivan Eifel, I. Velikanov published a book discussing secret research in fascist Germany into the spread of germs in the Paris subway system. The implication of Velikanov's book is that the Soviets may have been helping Germany with the research and design of biological weapons. Unquestionably there was considerable cooperation between the Soviet Union and Germany in the twenties and the early thirties on the development and testing of chemical weapons. This research was carried out in Shikhani, about 140 kilometers north of Saratov near the Volga River.[17] In any event, Eifel's book made quite a stir in Europe and the USSR.

Velikanov's direct boss was corps doctor M. I. Baranov, head of the Red Army Military Sanitary Administration. From 1933 to 1937, the institute studied the possibility of using aircraft and artillery to carry biological weapons; a special reconnaissance tank was built to diagnose biological weapons dispersed by an enemy. A universal gas mask was also designed to reliably protect people from both chemical and biological offensive agents. An experimental gas mask called B-3 was produced that met the proposed requirements, and a portable laboratory kit to identify types of biological contamination was designed.

An island (Vorozdenie or "Rebirth" Island) in the Aral Sea, and other sites as well, were already used at that time for various biological weapons experiments. Rebirth Island was a large, uninhabited place, used as a site for top-secret research. Only a few medical officers of the organization P.O. Box A-1968 knew about it.[18] Various biological weapons, including plague, anthrax, and even smallpox were tested on the island, once with tragic consequences.

In 1971, when smallpox had long been eradicated from the Soviet Union, a research ship studying plankton on the Aral Sea passed unknowingly within fifteen kilometers of the test sight; tragically, a smallpox bomb, using a powerful strain, had been exploded not long before. A young woman technician was on the ship's deck. Despite having been vaccinated, she came down with severe smallpox. She passed the disease to her brother, who had also been vaccinated, and he passed it on to eight other people. Three unvaccinated people, two of them infants, died of hemorrhagic smallpox, the deadliest form of the disease.[19] But this incident was not known for many years, even in Russia, or even to me.[20]

After a laboratory-caused plague outbreak in Moscow, research into agents of especially dangerous infections was discontinued in Vlasikha. The plague researcher A. L. Berlin (see chapter 3) had immunized laboratory animals with the live vaccine strain EV, and then apparently "challenged" them by spraying them with live, virulent plague bacteria. In the course of this experiment Berlin infected himself with plague. He traveled to Moscow while incubating the disease; soon after, he died of pneumonic plague. But several other people, including his barber, contracted the disease from him and also died. Fortunately, the outbreak Berlin unwittingly caused in Moscow was contained, thanks to appropriate anti-epidemic measures. Berlin's accident showed clearly that work on potential agents of biological weapons represents a danger to cities, and therefore requires very special safety conditions.

The military took Berlin's death as a warning: work with plague and anthrax shifted to the remote Kirov Military Institute soon thereafter. During or soon after the war the Kirov Military Institute mastered the technique of lyophilization—freeze-drying cells under vacuum pressure—of strains of the plague germ. Freeze-drying makes the plague bacteria much better able to withstand heat, cold, or the drying effects of the external environment. The development of biological weapons requires the ability to work with dry live bacteria—so the development of this technique was essential. Also at Kirov, Dr. N. N. Ginzburg[21] obtained a live anthrax vaccine, the STI strain, which is still in use in Russia to deal with the occasional cases and small outbreaks that occur from time to time.[22] The institute also was the first to develop a large-scale technology for preparing the live antiplague vaccine on the basis of the EV strain, which is still used in Russia as well.

Also, similar work, though connected with viruses and rickettsias[23] rather than bacteria, had begun at the Zagorsk Military Institute near Moscow. P. F. Zdrodovskij, a well-known scientist and a ranking official at the Zagorsk Institute, took an active part in the institute's researches. Many alumni of the Zagorsk Institute regard him as their teacher.

Like other scientists, Zdrodovskij had once been arrested, but by the 1940s and 1950s, despite his earlier incarceration, he had developed a close relationship with the Military Institute in Zagorsk. He worked with Zagorski scientists on rickettsia research. Unfortunately, it is impossible now to learn any details about the arrests of P. F. Zdrodovskij and other microbiologists in the thirties and the forties. These documents may be hidden in the archives of the KGB (now FSB), and are still kept secret. We also cannot know why Zdrodovskij and others, once imprisoned by the Soviet system, later went to work for it.

I do not know when biological weapons work began at the Military Institute in Sverdlovsk (now Ekaterinburg). We do know, however, that by the fifties they were already engaged in research with anthrax.

The historical reconstruction of the development of biological and other weapons in the Soviet Union in the pre- and postwar years is an almost insoluble task, since a lot of work was done by prisoners in so-called sharashki.[24] There were sharashki for experts in aircraft construction, atomic and radio research, and engineering as well as for microbiologists.

One of these imprisoned microbiologists was N. A. Gaiskij (see chapter 4) who discovered, together with Dr. Elbert, a live antitularemia vaccine. He was arrested in 1930 and sentenced to five years in prison under Article 58-1 (the usual charge of anti-Soviet propaganda). The questionnaire filled out by Gaiskij (it is kept at the Irkutsk Anti-Plague Institute) says that he was a tester at the Third Testing Laboratory of the Red Army. Like Zdrodovskij, he continued to work there after his release. L. A. Zilber, the founder of the virus theory of cancer, also spent a brief period (a year or two) in prison just before the war. One can only speculate about what he did in the prison sharashki.

It is very difficult to explain to readers in the West how scientists could be imprisoned for imaginary offenses, and then, once they were set free, remain in the same institutes doing the same work for the very government that had imprisoned them. Quite possibly, the scientists whose expertise was needed by the government would simply be arrested on obviously trumped-up charges and put to work on particular projects. Afterward, they may have been threatened or coerced in some manner to continue their research. I do not know. At such prisons or camps there were the requisite conditions for creative work, though they certainly could not be mistaken for a health resort! In any event, after discharge some experts, like Gaiskij and Zdrodovskij, went on to work at these camps voluntarily, as civilians.

In the period from 1950 to the 1960s, the direction and level of military research to develop biological weapons can be gleaned from Gen. N. I. Aleksandrov and Col. Nina Gefen's book.[25] Gen. P. N. Burgasov,[26] a deputy head of the Sverdlovsk Institute before his appointment as deputy health minister of the USSR in 1965, maintained a close relationship with Aleksandrov and Gefen at that time.

In addition to the military institutes in Kirov (Vyatka), Zagorsk (Sergiev Posad), and Sverdlovsk, there was also the 32nd Institute in St. Petersburg. Major General Gapochko, who was engaged in work on Problem No. 5, once headed this institute. When I was the director of the Plague-Control Institute in Rostov-on-Don, every year I would receive teams of inspectors who came to check on our work on Problem No. 5. These teams always included people from the 32nd Institute (including Colonel Goldin, whose first name I do not remember, but who was an expert in microbe diagnostics).[27] Despite the nominal involvement of the 32nd Institute in Problem No. 5, I have no doubt that their interest in biological weaponry went far beyond issues of biodefense.

In 1975 V. I. Ogarkov and K. G. Gapochko (General Ogarkov was the first chief of Biopreparat [see chapter 6]) published the book *Aerogenic Infections*, dealing with germs that can be spread through the air, or, in other words, transmitted directly from person to person through the respiratory route. The aerosol route is, of course, the chief route of infection in a biological weapons attack.[28] Last, I have to mention the book *The Inhalation Vaccination Method*, by Gen. V. A. Lebedinskij.[29] All of these scientists were both teachers and ideologists at the military institutes, and deeply involved in biological weapons research. These books are subtle but tangible evidence of their profound involvement with the research and design of biological weapons.

But the ideologue par excellence of Soviet bioweapons research from the 1950s to the 1980s was Col. Gen. E. I. Smirnov, who knew Stalin personally. During the war he headed the Army Medical Service. Thereafter he served as minister of health until 1953. The Doctors' Plot episode, and the subsequent persecution of Jewish doctors, took place during Smirnov's tenure. Until his death in 1985, Smirnov was chief of the 15th Directorate of the Ministry of Defense (Organization P.O. Box A-1968). Smirnov's activities before the end of the war are described in his memoirs *War and Military Medicine*.[30]

In December 1949 Soviet scientists showed great interest in the case against Japanese war criminals in Khabarovsk, who, as I noted previously, had developed various biological weapons and tested them on prisoners—whom they called "logs"—in occupied Manchuria during World War II.[31] N. N. Zhukov-Verezhnikov, a member and vice-president of the prestigious USSR Academy of Medical Science, appeared at the trial in the role of medical expert. The relevant proceedings were published. I came across the record quite by chance.

I became interested in the Japanese case a good deal later. After 1973, when I had been appointed as deputy chair of the secret Interagency Science and Technology Council on Molecular Biology and Genetics—the "brain center" of the entire Soviet biological weapons apparatus—I needed to see this material in order to discover certain technical details of Japanese work on the large-scale culture of bacteria and fleas. I believed that the abundant experience of the Japanese "Unit 731" with the "large-scale culture of bacteria and fleas" might be useful for us. I naturally approached Zhukov-Verezhnikov, but in spite of our good relationship I couldn't get any information out of him. I didn't even succeed in learning where to find the technical details. Obviously they still remained secret for some reason. Zhukov-Verezhnikov, despite our long acquaintance (I had known him from the early fifties and worked closely with him in the Nukus cholera epidemic), was a reticent man, and I was in no position to explain to him exactly why I needed the information. Both the place and the purpose of my work, even the nature of my post itself, were top secret. The veil of secrecy that has been thrown here over the proceedings of the case in Khabarovsk therefore seems suspicious in retrospect.

Also linked with the name of Zhukov-Verezhnikov, who by then was the deputy minister of health, is another murky episode in the history of biological weapons. Zhukov-Verezhnikov was a member of the international commission to investigate the supposed facts of biological warfare by the U.S. Army in China and Korea in 1952. I learned the details from a book given to me in 1957 in Beijing.[32] However, so far as I can learn from the Stockholm International Peace Research Institute (SIPRI)[33] material and from remarks by my Chinese colleague Di Shu-li, the results of the commission's work were, to put it mildly, "exaggerated." Nevertheless, even though there is no reason to think that the United States actually used biological weapons in Korea, research into the development of biological weapons was certainly carried on for a number of years since the 1930s both in the United States (Fort Detrick in Maryland) and at Porton Down in Britain. The question is merely whether this biological weapons research was ever actually used for anything, except by the Japanese.

Soviet scientists felt at the time that the well-known work by T. W.

Burrows and G. A. Bacon[34] on the determinants of virulence of the plague agent, which incidentally still retains its importance, was a spin-off from the British military program. (I heard this often from Zhukov-Verezhnikov.)

Our allies in the Warsaw Pact also showed interest in biological weapons, as may be inferred from the book *Biologicheskaya voyna* (Biological Warfare), by Polish authors T. Rojniatovski and Z. Zholtovskiy.[35] Based on this book, it seems that our allies were also involved in the Soviet biological warfare program. Although the book draws on foreign sources, it was written by people who knew their subject. A dilettante could not have written a book like this, with its thorough military assessment of various agents (bacteria, viruses, and toxins), the means to produce and disseminate them, and the general strategy and tactics of biological war. It is interesting that I never saw such openly explicit books written by Russian authors. There were only the "indirect" presentations, like the books by Ogarkov and Gapochko or Lebedinskij, which I discussed above.

Furthermore, Polish authors T. Rojniatovski and Z. Zholtovskiy discuss Hitler's biological weapons development program. It is quite possible that some of these documents ended up in the German Democratic Republic and were available to the Soviet military (according to the authors, some of the documents and even some specialists ended up in America after the war).

There is also further evidence that the Soviet Union's Warsaw Pact allies were involved in the development and testing of biological weapons. During the 1960s, when the construction of an aerosol building was being planned at the Rostov Antiplague Institute, for purposes of Problem No. 5, as we were told, some Czechs worked as planners for four years (1965–1969). They were also supposed to deliver equipment to us for this building. With Maj. Gen. N. I. Nikolaev (he was then head of the Mikrobe Institute in Saratov) I was in Czechoslovakia three times and discussed there all the relevant details for this project. But the project was very expensive and ultimately our Ministry of Health canceled the treaty with the Czechs.

I heard later, though, from Gen. V. Lebedinskij that the Czechs were building a similar block for their own needs. Pehaps this project was done under the Warsaw Pact for the testing of offensive biological weapons, but not for experiments in protection. It was too elaborate a structure for biodefense work. In any event, Lebedinskij's visit to Czechoslovakia in the late sixties or early seventies indicates a direct connection between Czech medical officers and those of the Soviet Union. Interestingly, all our aerosol equipment was manufactured at the Chirana Medical Equipment Plant in Brno, Czechoslovakia.

The Soviet bioweapons program itself vigorously pursued the development and even the testing of biological weapons on Rebirth Island in the Aral Sea until the early or mid-1960s, but then the effort began to wind down.[36] This slowdown coincided with the cessation of work on building the aerosol block in Rostov. As I learned from our apologist for biological warfare, Colonel General Smirnov, the reason was that, since we already possessed nuclear weapons, someone high up questioned the usefulness of any further development of biological weapons. The use of infectious disease agents for military purposes would make it difficult to avoid some "reverse effect," i.e., the transfer of epidemics from the enemy to the people using them. Furthermore, military use of infectious agents would involve employing unaltered strains of bacteria and viruses, which were unable to meet the tactical and technical requirements for weaponry. Later on, of course, we succeeded in developing altered bacterial and viral strains, which differed from natural strains by the introduction of new and often unusual characteristics such as resistance to many antibiotics and other treatments (in the case of bacteria), or modified antigenic structures (to evade natural or vaccine-induced immunity), or enhanced survivability in the aerosol form. These strains were more predictable in their effects, and more likely to meet the tactical and technical requirements for weapons than the unaltered strains developed early in the program.

But Smirnov felt that, one way or another, any winding-down of these activities would inflict serious damage on our defense capability. And despite all the discussion in the sixties and early seventies, research and development of biological weapons never actually ceased. Smirnov's faith in the essential military value of biological weapons appears to have been justified, because we had indeed scored some successes, mainly in the development of reliable technologies and equipment. I suggest this because, when the former head of the Research Institute of Epidemiology and Hygiene at Kirov, Gen. N. L. Nikolaev, was appointed director of the institute in Saratov, he brought with him some of his scientists as associates. They were able to refurbish the entire production system for the antiplague vaccine, which used to be extremely outmoded.

The studies I have mentioned on the use of dry aerosols for aerogenic vaccination, as well as the book *Aerogenic Infections,* remove all doubt that the Rosebury doctrine was fully accepted by our military people, and also reflect their progress in this field. This raises the question of why these people spoke only in whispers about activities that, in the opinion of specialists, were supposed to enhance the defense capability of the Soviet Union, while, at the same time, they loudly asserted that "the science of biological warfare was science stood on its head and a gross perversion of science."[37] If these activities were undertaken for defense purposes, would it not be better, without giving away any secrets, to inform the world community of our successes (as had been the case with Voroshilov's open threats of 1938)—without having recourse to any ideological contortions? It is perfectly well known that the fact of our developing an atom bomb, which wasn't a secret, played a positive part in deterring the "aggressors"!

In any event, strange and perhaps counterproductive as it seems today, the Soviet biological weapons program was plunged into the deepest possible obscurity since its inception: all that we have of Soviet work in the early days of the program is circumstantial evidence, which I have outlined in this chapter. Only after former president Boris Yeltsin openly issued a decree in 1992 shutting down the program and prohibiting the development and testing in Russia of all biological weapons did the former Soviet program begin to emerge from the shadows.

In the following chapters I discuss my own experience in the field, which I no longer wish to keep secret.

NOTES

1. T. Rozhnyatovskiy and Z. Zholtovskiy, *Biologicheskaya voina* (Biological Warfare) (Moscow: Foreign Literature Publishers, 1959).

2. Charles Nicolle, *Birth, Life, and Death of Infectious Diseases* (in Russian) (Moscow: State Publisher House of Biological and Medical Literature, 1937).

3. Ibid.

4. Ibid. Also, see the recent book by American historian Elizabeth Fenn, *Pox Americana: The Great Smallpox Epidemic of 1775–82* (New York: Hill and Wang, 2001).

5. See below, p. 131.

6. See *The Problem of Chemical and Biological Warfare*, vol. 1, *The Rise of CB Weapons* (Stockholm: Almqvist & Wiksell, 1971).

7. The Research Institute of Epidemiology and Hygiene at Kirov was created in 1928.

8. The Kronschtadt "plague station" was actually a sea fortress. Kronschtadt is a town and a port on Kotlin Island in the Gulf of Finland (about 30 km. from St. Petersburg).

9. This article was based on a decree from the Supreme Commission on Measures for Prevention and Combating Plague Infections of August 22, 1899.

10. This situation of isolating research on highly contagious pathogens to remote locations has parallels in the West. For instance, the U.S. government exploited Utah's western desert as a site for its World War II biological and chemical weapons research. The intent was that placing work in such a remote and inaccessible area (now known as Dugway Proving Ground) would lessen prospects for accidental escape of any toxic substances. Such hopes were dashed by the 1968 accidental release of the chemical warfare agent "VX," causing the death of thousands of sheep on ranch lands in the nearby Skull Valley. In the early 1950s, the U.S. Department of Agriculture began using Plum Island as a center for research on highly contagious animal pathogens. Plum Island, off the northeast coast of New York's Long Island, is the only U.S. site permitted to store and study foot-and-mouth disease virus, a highly contagious pathogen whose release on the continental United States could have devastating consequences for the livestock industry, but would pose little threat to humans. (Ben Garrett)

11. Presumably he was referring to the Geneva Protocol of 1925.

12. I am quoting from *The Problem of Chemical and Biological Warfare*, vol. 1, p. 287.

13. Kanatjan Alibekov, now known as Ken Alibek, was the former head of Biopreparat who defected to the United States in 1992, soon after the fall of the Soviet Union. He has written (with Steven Handelman) an account of his work in the Soviet bioweapons apparatus called *Biohazard* (New York: Random House, 1999).

14. V. I. Velikanov, *Sudby ludskie* (People's Fates) (Moscow, 1998).

15. Anaerobic bacteria are bacteria that grow only in the absence of oxygen.

16. *Clostridium botulinim* produces one of the most powerful toxins known.

17. See *The Problem of Chemical and Biological Warfare*, vol. 1, pp. 284–87. From 1922 until 1933, the governments of Germany and Russia cooperated in many spheres. This cooperation was especially brisk in the military arena, with the Soviets providing the Germans military training and weaponry forbidden to them under the peace treaties ending the First World War. This cooperation extended to tanks, military aviation, and chemical weapons. A major site for developing and testing these chemical weapons was in Shikhany, less than a hundred miles to the northeast of Saratov, where the author was living. The story of the Russo-German cooperation is detailed in Yuri Dyakov and Tatyana Bushyeva, *The Red Army and the Wehrmacht: How the Soviets Militarized Germany, 1922–33, and Paved the Way for Fascism* (Amherst, N.Y.: Prometheus Books, 1995). (Ben Garrett)

18. Intensive irrigation for cotton farming in the former Kazakh and Uzbek Soviet Socialist Republics diverted so much of the brackish Aral Sea waters that much of the lake has dried up, and Rebirth Island is now a peninsula. Therefore, there is relatively easy access to Vorozdenie and its buried tons of pathogens, including anthrax and plague. Rebirth Island is now the territory of Kazakhstan, not Russia. There is now a joint project between Kazakhstan and the United States to clean up Rebirth Island.

19. This account is based on the analysis of the American biodefense specialist Alan P. Zelicoff of Sandia National Laboratories, Albuquerque, New Mexico. Alan P. Zelicoff, M.D., "An Epidemiological Analysis of the 1971 Smallpox Outbreak in Aralsk, Kazakhstan," in Jonathan B. Tucker and Raymond A. Zilinskas, eds., *The 1971 Smallpox Epidemic in Aralsk, Kazakhstan, and the Soviet Biological Warfare Program*, CNS Occasional Papers: #9 (2002).

20. The details of this accident have only recently been brought to light. General Pyotr Burgasov, whom we met in the last chapter, described the incident in a November 13, 2001, interview in *Moscow News*, in which he suggested that anthrax is really not such an effective biological weapon and that terrorists might

want to try smallpox instead: "Smallpox—that's a real biological weapon," said Burgasov.

21. See Glossary of Names.

22. Anthrax outbreaks among livestock occur worldwide, in the United States and other Western countries as well as in many parts of Asia; the Sterne vaccine strain used to control these outbreaks is similar to STI, but in the West it is only used on animals.

23. Rickettsias are a kind of bacteria. Some of them are causal agents of spotted fever, including Rocky Mountain Spotted Fever, typhus, and other tick-borne diseases. Rickettsias are obligate intracellular bacteria, which means that, like viruses, they can grow only inside living cells.

24. The system was described well in A. Solzhenitsin, *V Kruge Pervom* (The First Circle), trans. Thomas P. Whitney (New York: Harper and Row, 1968).

25. The title of Aleksandrov and Geffen's book, translated from Russian, is: *Active Specific Prevention of Infectious Diseases and Ways of Its Improvement.* See the bibliography at the end of the book.

26. Burgasov, as we see below, was deeply implicated in the official Soviet cover-up of the Sverdlovsk accidental anthrax release in 1979.

27. Colonel Goldin is a collaborator of the 32nd Institute in St. Peterburg. His chief was General Gapochko.

28. V. I. Ogarkov and K. G. Gapochko, *Aerogennye infektsii* (Aerogenic Infections) (Moscow: Meditsina, 1975).

29. V. A. Lebedinskij, *Ingalyatsionnyiy (aerogennyiy) metod vaktsinatsii* (The Inhalant [Acrogenic] Method of Vaccination) (Moscow: Meditsina, 1971).

30. "Voyna i voennaya meditsina" (Moscow, 1979).

31. A full account of the activities of the infamous Unit 731 may be found in Sheldon Harris, *Factories of Death: Japanese Biological Warfare, 1932–45, and the American Coverup* (Routledge, 2001).

32. Report of the International Scientific Commission to investigate the facts of biological warfare in Korea and China. Beijing 1952.

33. Almqvist & Wiksell, 1971, vol. 5, pp. 238–58.

34. Burrows and Bacon did not publish a book on this subject: their work appeared in a series of articles in various scientific journals. Also, I had the opportunity to get to know Burrows both in Moscow, in 1966, and then again in Malmo, Sweden, in 1972.

35. Moscow: Foreign Literature Publishers, 1959.

36. V. Umnov, "Posle 20 let molchaniya sovetskie mikroby zagovorili"

(After the 20 years of Silence Soviet Microbes Begin to Speak). *Komsomolskaya pravda*, no. 80, April 30, 1992. This article played a significant role in opening the secret history of Soviet biological weapons to the general public in Russia.

37. M. I. Rubinshtejn, *Burzhuasnaya nauka i tekhnika na sluzhbe amerikanskogo imperializm* (Bourgeois Science and Technology in the Service of American Imperialism) (Moscow: USSR Academy of Sciences, 1951), p. 285.

CHAPTER 8

RENAISSANCE

(THE NEW STAGE)

If you would have peace, prepare for war.

Vegetius[1]

S ince the early 1960s many Russian scientists realized that, compared to the real progress made by Western molecular biologists and geneticists, we hadn't achieved very much. We were hobbled by the decades-old Stalinist devotion to the dubious "genetics" of Lysenko, an intellectual totalitarianism that only began to fade in the early sixties. We also recognized the military utility of the new advances in molecular biology and genetics that had been made by scientists in the West. The VPK (Military Industrial Commission) began to take an interest in these fields with a certain delicate touch: much of the actual work was carried out by civilian scientists, and the creation of the new Problem Ferment, the code name for the entire Soviet biological weapons program, beyond a doubt originated with these civilian scientists.

Nevertheless, the military used every possible method and occasion to learn about the most important achievements at our institutes, in particular at the plague-control institutes like Rostov, Saratov, and Irkutsk. Military people were among the members of all commissions of the Ministry of Health. These commissions inspected all our institutes; I do not think these inspections were carried out for defensive purposes.

The organization of the Military Industrial Commission itself was far removed from science; its directors and staff were mainly officials, as well as military men. In order to devise future defense industry plans, therefore, the commission used our civilian scientists to gather information, which they used to recheck the data derived from intelligence.

First, these civilian scientists gathered and analyzed all published data on molecular biology and genetics and presented a summary of this data to the commission. Second, they visited the most important conferences abroad and helped the commission to analyze that information. Finally, they helped the commission analyze material submitted by the Main Intelligence Service. Taking part as active informers were the famous scientists Yu. A. Ovchinnikov, G. K. Skryabin, and particularly, V. M. Zhdanov, a noted virologist who had initially proposed the world-wide campaign to eradicate smallpox[2] and who had won the high regard of many Western scientists. He was also responsible for many original ideas in virology. Zhdanov had traveled abroad extensively and had often received foreign guests; this sophisticated, worldly man was therefore a great asset to the military. Every time any of these scientists took a trip abroad, they had to present a report to the commission. Most probably, agreeing to gather such information was an indispensable condition for any foreign travel.

In their reports to the commission, Ovchinnikov, Skryabin, and Zhdanov were each pursuing self-seeking goals—their desire for rewards and academic posts. With the exception of Zhdanov, they lacked serious interest in Problem Ferment. And they were indeed rewarded for their efforts: Such scientists as Ovchinnikov, Skryabin, and perhaps Baev obtained many high state rewards and posts in the Academy of Sciences during this period. No one questioned these promotions, so far as I know: most of their fellows perceived these promotions as normal everyday occurrences, as recognition of their scientific merits.

All discussions with these scientists were held in secret in the Military Industrial Commission and were not even disclosed to heads of the academies (of science and of medical science) or the relevant departments. These discussions focused on the necessity of matching Western results in the fields of molecular biology and genetics in order to develop an effective, sophisticated bioweapons program; they also addressed the creation of a special scientific and practical organization to implement this program.

A special commission was set up as a result of these discussions. I don't know its composition, but according to Zhdanov it included himself and O. V. Baroyan from the Academy of Medical Sciences. It also included members of the Military Industrial Commission and the Central Committee of the Communist Party. This commission was allowed into "the holy of holies"—Smirnov's "domain"—to acquaint themselves with the state of affairs in the development of biological weapons. It was extraordinary for civilian scientists to be admitted to these military institutes: as a rule, no civilians were ever allowed there. But their presence was deemed necessary: one of the tasks required of these civilian scientists was a scientific estimation of the results already obtained by the military and of what changes would be necessary for the preparation of a new, recharged biological weapons program. Ovchinnikov, Zhdanov, and others tried to prove to the military that all their previous work had been done in the old, Lysenko-tainted way, without the application of modern techniques of molecular biology and genetics, without the modification of strains that would produce stable, workable biological weapons.

The secret meetings had their effect: The Party and the government issued orders identifying what had gone wrong in earlier research and how those failures might be remedied. The government intended "to catch up and leave behind" any "potential enemies"—that is to say, the United States and China.

One directive set up a new "civilian" system for the microbiological industry under the aegis of the Main Directorate of the Microbiological Industry at the USSR Council of Ministers. It paralleled the work going on in the military labs such as Kirov and Zagorsk. The key task for this system was the treatment of fundamental problems in molecular biology, genetics, and advanced technology, in order to solve certain applied problems for the military—the production of viable, suitably modified strains, the production of these strains, and their conversion into forms appropriate for dispersal—in other words, to establish a firm scientific foundation for the development of offensive biological weapons. This directive also provided for the construction of modern scientific centers and institutes, as well as for the transfer, within six months, of one of the antiplague institutes to Glavmikrobioprom for work on the study of pathogenic bacteria for offensive purposes. (This institute—the antiplague institute of Volgograd—had originally been a plague-control station, but had been converted to a full-fledged antiplague institute in 1970, though it never actually engaged in plague research. Instead, Volgograd was actually dedicated in work on biological defense—Problem No. 5.) Incidentally, owing to the opposition of the minister of health, this part of the order was not implemented; I had insisted on the transfer because Glavmikrobioprom had no research institutes for work on pathogens, but Burnazyan refused to allow it. He insisted instead on the rapid building of new institutes for Glamikrobioprom. Apparently, he considered this a very important political matter, and outside my jurisdiction. This provoked a quarrel between Burnazyan and me, which I deeply regret.

One of the first directives, in 1971, of this new system under Glamikrobioprom proposed to earmark large sums of foreign currency for the procurement of requisite equipment, reagents, and copies of relevant literature. Foreign currency was necessary if we were to go abroad to purchase equipment, reagents for molecular biology and genetic research, and corresponding scientific literature. We did not have the appropriate materials or scientific literature in our own country[3] and to procure them abroad took money, valued foreign currency, which only the government could supply. We lacked even the simplest of the especially pure formulations, such as sodium chloride, needed to conduct experiments in molecular biology. And in order to "catch up to and surpass" other nations, we had to master the work their scientists had already done in molecular biology, genetics, biochemistry, and microbiology. Gaining access to Western scientific publications was therefore essential.

Finally, the directive set up an Interagency Science and Technology Council for Molecular Biology and Genetics (the Russian abbreviation is MNTS) under Glavmikrobioprom. This became the "brain center" of the new organization, and therefore of the entire Soviet biological weapons apparatus. The functions of the Interagency Council included, in particular, the design and drafting of scientific programs, the coordination of the activities of all the participating authorities (the USSR Academy of Science, the Ministry of Health, the Ministry of Defense, and even the Ministry of Agriculture), and the distribution of currency allocations. A special department, under the council's deputy chairman, was designated to carry out the council's preparatory work. This special department prepared documents for the council: corrections of plans for the institutes and factories of Biopreparat and other departments (though not for the laboratories run by the Ministry of Defense), estimations of the scientific and technological results of their work, and so on. All of these documents were then considered by the council and approved by the Higher Certification Committee. Furthermore, the special department prepared drafts of letters for the chairman of the Interagency Council to be sent to the government and the Central Committee of the Communist Party.

Zhdanov was appointed chairman of the Interagency Council, while I was appointed his deputy and head of the special department of the council. Besides us, the council also included Maj. Gen. Lebedinskij (from the Ministry of Defense); A. A. Skladnev and S. I. Alikhanyan (Glavmikrobioprom); Burnazyan (the Ministry of Health), Ovchinnikov, Skryabin, and A. A. Baev (the Academy of Science). Furthermore, there was a representative of the USSR Ministry of Agriculture, but I have forgotten who it was. Most of these people, in truth, were secondary figures with no direct knowledge of, or relation to, the research and design of biological weapons. The inappropriate nature of many of these appointments was one of the reasons for the reorganization of the council in 1975.

R. V. Petrov, the distinguished immunologist, now a vice-president of the Russian Academy of Science, also had, at first, some connection with the council, but he was soon removed as "redundant." He then made out that he couldn't understand what was expected of him, or he didn't want to understand. Petrov was close to Burnazyan, and perhaps had learned something about biological weapons from him. On the other hand, Petrov was expert only in general immunology: he never worked directly on microbiology or virology, so his background and expertise were not quite appropriate for our problems. Furthermore, at about this time, Burnyazan had created a new institute for Petrov, the Institute of Clinical Immunology, so his removal from the council could not have been much of a loss for him.

The Politburo of the Communist Part of the Soviet Socialist Union and the USSR Council of Ministers confirmed our appointments, which were also approved and signed personally by General Secretary of Central Committee of C.P.S.U. Leonid Brezhnev and Premier Alexei Kosygin.

V. D. Belyaev always attended sessions of the council (where Zhdanov usually yielded the chair to him as the principal personality), together with representatives of the party, the Military Industrial Commission, and of necessity the head of a directorate of the KGB. All of these orders and decisions, including the construction of new installations, were implemented by still another secret organization, established by Glavmikrobioprom, and known officially as P.O. Box A-1063 or Biopreparat.[4] It was also called, informally, "Ogarkov's organization," since V. Ogarkov, deputy head of Glavmikrobioprom, and ex-chief of the Military Research Institute in Sverdlovsk, was tapped to run it.[5]

At the beginning of the seventies, after the signing of the 1972 Biological Toxin and Weapons Convention,[6] the new Biopreparat or Organization P.O. A-1063 assumed leadership of the Soviet biological weapons program. Biopreparat was established under the auspices of Glavmikrobioprom, which was in turn under the aegis of the USSR Council of Ministers. The Interagency Scientific and Technical Council for Molecular Biology and Genetics, described above, became the nerve center of the entire system. Biopreparat's research efforts focused on the development of bacteria resistant to antibiotics, and on antigen-altered bacteria and viruses, as well as upon improving the survival of microorganisms in the environment. Furthermore, the system was required to create an industrial base for the large-scale production of biological weapons.

At this point, it is useful to remember that the new program did not emerge de novo, nor was it entirely unprecedented. As I have shown in the preceding chapter, the Soviet biological weapons program began to develop in the very earliest years of the state. And, until 1969, the United States had an active biological weapons program quite similar to the Soviet one, as we understand from the 1967 account by the American scholar Milton Leitenberg.[7] According to Leitenberg, there are a number of criteria that make a biological agent appropriate as a weapon, including the capacity to incapacitate or kill; the ability to produce the agent economically and disperse it effectively; the ability to retain its infectivity throughout processing, transportation, dispersal, and other attributes. Eventually the U.S. program, fearing "blowback" or the ability of an agent to infect the nation that dispersed it, switched to supposedly nonlethal agents such as tularemia and Q-fever, in preference to anthrax, smallpox, and plague.[8] (Tularemia, in fact, can indeed cause a fatal infection, especially its "American" varieties.) These weapons were intended mostly as tactical weapons, meant for launching at enemy troops on the battlefield. The Soviet program, however, preferred to develop strategic weapons (meant to be launched deep into enemy territory), and these were meant to be lethal.

Leitenberg's description was written well before the era of molecular biology and genetically engineered organisms. Nevertheless, it is a full reflection of the similarity between the two programs! I can only add here that both programs used aerosols as a principal vehicle for biological weapons. I think that the American reader, reading this, should decide for himself who were the "pioneers" in this case.

The first scientific aim of Biopreparat's work, which began in the seventies, was the development of pathogenic bacteria resistant to many antibiotics ("polyresistant" strains); this would prevent the treatment of the corresponding infection. Biopreparat's second aim was to obtain strains of bacteria and viruses with modified antigenic structures, in order to overcome the specific immunity to a particular infection. In other words, if a means could be found to alter the way a particular virus or bacterium appears to the human immune system, natural or vaccine-induced immunity could be overcome. Thus, the human organism might be made more sensitive to infections. Third, Biopreparat had to find methods to increase the survival of microorganisms in aerosol form to intensify their efficacy. The new techniques of molecular biology were required to achieve these aims.

From the purely technical perspective, however, Biopreparat's tasks were far more various, ranging from the design and engineering of institutes and munitions factories to the development of new means of large-scale biological weapons production. Many leading members of Biopreparat were specialists commandeered by Colonel General Smirnov and enlisted into active military service. Since everyone went about in civilian clothes it was difficult to tell which among us were "servicemen."

The Special Department of the Interagency Council,[9] which I headed, also had an independent "post-box number": Organization P.O. Box A-3092. The department of the council was composed of assistants to members of the Interagency Council who were supposed to prepare the materials for their chiefs. These materials consisted of the analysis of different documents, including plans and accounts of scientific work.

Col. L. A. Klyucharev[10] was appointed my deputy. He was a very intelligent and decent man, eventually promoted to general and head of the science department at Biopreparat. Working in the Organization P.O. Box A-3092, I had to associate daily with other "Old Petersburgers"[11] besides Klyucharev: V. A. Lebedinskij; Maj. Gen. I. P. Ashmarin, the "creator of ideas" in Smirnov's "domain," an erudite man responsible for the subsequent synthesizing of mammalian neuropeptide genes by microorganisms, and for the creation of "Factor," one of the most significant divisions of the Soviet bioweapons program;[12] Maj. Gen. V. N. Pautov (director of the RIEH); V. I. Ogarkov; and D. V. Vinogradov-Volzhinskiy, whom we shall meet again. They were all from good families and were alumni of the Military Academy of Medicine. Klyucharev was slightly older than most of them, and the only one to have served at the front during the war with Germany (1941–1945) as a medical assistant. Everyone thought it quite normal for him to be working in the highly specific Smirnov system—the system, that is to say, responsible for the creation and development of biological weapons. Obviously it couldn't have been otherwise. He never had any doubts or hesitations, as befits a soldier. Klyucharev was a military man his entire life, and he liked his work. I think he could not have imagined any other existence. On the other hand, I was the only real civilian among them, but then I had very little time for reflection, perhaps due to my vaulting pride that I was one of their circle, although I should add at once that I never became one of them. The supposedly classless society of the Soviet Union had castes of its own.

I had no doubts as to the country's need for our work, as it was directed toward the solution of strictly scientific problems. The creation and development of biological weapons is also a scientific problem; without advances in molecular biology and genetics, which was my chief concern as deputy chair of the Interagency Council, the scientific problems facing bioweapons designers could not be solved.

I greatly appreciated my transfer into the Glavmikrobioprom system. It was only later that doubts of a moral nature arose when I was working in the Organization P.O. Box V-8724, the All-Union Institute of Applied Microbiology at Obolensk, and a laboratory run directly by the military. There I heard about what was "really happening": the weaponized strains, the cultivation and lyophilization of bacteria on a large scale, the chambers for aerosol infection of monkeys, the equipment for loading of missiles, and so on.[13]

But at first (1973–1975) I had nothing to do directly with the practical aspects of weapons development. V. Zhdanov and I were civilians. We were involved in the scientific realm, not the applied aspects of bioweapons work. We thought in terms of creating genetically modified strains, not of creating bombs and missiles filled with weaponized agent. We knew how to create these genetically modified agents in principle. But we thought of them as strains, not as weapons.

But as I learned later, the production of strains of microorganisms with specified properties, in which I was directly engaged, was merely a prelude to the real business. Strains of microorganisms, no matter how they are modified, are not in themselves actual weapons. The development and production of such strains are merely the prelude and the necessary condition to the development of a biological weapon. Since I was involved in solving the problem of how to develop antibiotic-resistant and other modified strains I understood of course that they might have a dual purpose. On the one hand, this work helped us to learn some fundamental aspects of bacterial evolution. But, on the other hand, the military significance of these strains was clear. It is important to stress that most of my young colleagues in the Moscow laboratory, and even later at the military bioweapons laboratory at Obolensk, took a great interest in fundamental research and seldom thought about their applied military significance. Some of the young workers at Obolensk were not fully devoted to Problem Ferment, and some even worked according to a "legend" or cover story. Even I myself often forgot about biological weapons, and published a whole series of articles on bacterial evolution.

For what it is worth, I have never taken seriously any arguments for the superior humanity of one type of weapon over another. To me, any weapon is inhumane, and biological weapons are no more so than guns or tanks or bombs. As has now become evident, more than two-thirds of the industrial enterprises and countless research institutes at that time came under the aegis of the Military Industrial Commission. It was inevitable in 1990 for a technological academy to be formed that united all those scientists who had decided to turn former major achievements in developing offensive and other weapons to peaceful purposes. Like myself, these scientists, regardless of their own convictions and attitudes to the existing system, had done their duty. It is only now, after a lot has changed, that the answer to the age-old Russian question "What is to be done?"[14] appears simple and unambiguous. We have to give up all of our achievements in molecular biology and genetic modification, or find another way to do science!

But matters were not so simple years ago. There was only one way to "do science" then—the Party's way. Scientists like myself faced essentially the same moral question as that concerning membership in the Communist Party. Many people managed to escape the temptations that accompanied being a Party member, and likewise there were a few scientists who resisted the desire to do important scientific work because that work was used to create bioweapons, but only a few of them achieved anything in life or got a chance to work in their professional fields. I am afraid that these thoughts will cause us to analyze the past and seek to justify our actions for a long time.

The whole activity of the new organization Biopreparat was kept a close secret, and the actual "problem"—the development of biological weapons—had a special code name, "Ferment" (abbreviated to "F"). It was not so much the nature of the problem, but the fact that we were actually working on biological weapons research, and the existence of the organizations responsible for this research, that were kept secret. Program "Ferment" was, therefore, termed a "Top Secret Regime." The Soviet Union had just at that time subscribed to the International Convention of 1972 banning the development, production, and stockpiling of biological weapons (we ratified it in 1975), so the secrecy was understandable. The entire program was built, as it were, in the shadow of the Biological Weapons Convention: these new organizations, the Interagency Scientific and Technical Council for Molecular Biology and Genetics (MNTS), as well as Biopreparat, were created just before and during signing of the convention (in 1972–1973). In order to conceal the top-secret regime and to "sanitize" all activities, there were two "cordons": public legends that cloaked the truth. These were (1) Problem No. 5—i.e., the development of means of protection from biological weapons, or research that was supposed to be purely defensive in nature, and (2) a public order by the party and government to enhance the development of molecular biology and genetics. The meaning of this order was clear: to let the nation and the world know that we had at last awakened and resolved to overcome our backwardness in this field.

To provide some legitimate underpinning, another council was set up, this one quite open. It was an "Interdepartmental" Council at the USSR Academy of Science, and its chairman was the same Ovchinnikov who also served on the secret Interagency Council. He had by then become one of the vice-presidents of the USSR Academy of Science. This should come as no surprise, as Ovchinnikov, the youngest academician in the country, was excellent at finding his way about the Soviet and party hierarchy. He was a good organizer, and more an astute politician than a scientist. He had access to the highest places, and was distinguished with all imaginable decorations.

I shall have many occasions to revert to the subject of the secret regime. It is worth focusing here on one G. I. Dorogov, who was the "heart" and "soul" of service to the regime at Biopreparat (P.O. Box A-1063). I don't know what the other counterintelligence experts of his rank were like, but Dorogov was a highly unusual character. He was just above average height with a large bald spot, very cheerful and straightforward at first glance. He was always compliant and modest toward his superiors, but by means of jokes and witticisms he knew how to talk to them and show that he was right. He usually came across as though he were seeking advice or requesting help in disentangling some set of affairs. Naturally his superiors gave the advice he desired and were duly "helpful." At the same time he could be very harsh and distant with his subordinates. Still, he had his positive aspects: he didn't fly off the handle, and would try to get to the bottom of any subject that was unfamiliar to him (I am thinking of biology). All of his working principles had been formed during his time in the Ministry of Medium Engineering Industry—the code name for the Ministry of Atomic Industry—back in the1950s, where he had gained his knowledge. Since Belyaev didn't sign a single piece of paper without Dorogov's approval, I had to deal with him frequently. These encounters were not always pleasant. His chummy attitude and condescending use of familiar forms of address left a nasty taste in many a mouth. Communist Party hacks—those counterintelligence agents responsible for secrecy in other departments of Biopreparat—largely imitated Dorogov. All documents concerning Problem Ferment were always transported with armed escort in special vehicles. The guards followed us everywhere; they left us only in the Ministry of Defense or the Kremlin (where they waited for us outside the Spasskij Gate).

Apart from the limitations on open publication, official strictness also affected our foreign travel. Throughout my time in the Interagency Council, I managed to travel abroad only twice, into socialist countries, and even then official representatives accompanied me. On one occasion this representative was Dorogov's deputy, whom I was obliged to introduce as my "senior scientific colleague." Luckily he was a candidate of science and could follow the general outline of my lecture.

Travel documents were made out for me fairly often, though all of these trips were canceled literally at the last minute. For example, in 1978, on the eve of my departure for Munich to attend an International Congress on Microbiology, my traveling expenses and ticket had to be surrendered right there in the street (it was after working hours) to a KGB man who was accompanying the delegation. Meanwhile, for some reason, the restrictions of foreign travel seemed to produce little effect on Ovchinnikov, Skryabin, or Baev, although they were likewise "under security."

There were three main reasons for the restrictions on my travel. First, my peculiar post as a head of the Special Department of the Interagency Council, as well as my deputy chairmanship of the council, made it risky, from the government's point of view, to allow me to travel abroad: I knew too much. Second, I was known abroad as an expert in dangerous infections. Third, there was a perennial atmosphere of universal distrust at Biopreparat, which, as it turns out, was more than justified. It was, indeed, dangerous to allow Biopreparat collaborators to travel abroad. In 1989, Vladimir Pasechnik, head of the Institute for Ultrapure Preparations in Leningrad, a Biopreparat-run institute, defected to England.* He was the first to reveal the secrets of Problem Ferment to the West. In 1992, he was followed by Kanatjan Alibekov (Ken Alibek), the deputy director of Biopreparat, who defected to America. But that was much later. Early on, very few scientists were permitted to travel: the relative freedom given Ovchinnikov, Baev, and Skryabin to travel abroad was granted only because of their positions at the Academy of Sciences. These restrictions on trips abroad caused me endless vexation, not to mention embarrassment, when I had to think up some reason for turning down pressing and very tempting invitations from my foreign colleagues. (I had either "broken" my leg, "caught' something, or I had to plead "family problems.")

Initially, P.O. Box A-1063 and my department were in Lestev Street on the premises of Glavmikrobioprom, but in the autumn of 1973 we moved to a building that until then had belonged to the Academy of

---

*I discuss this in chapter 18.

Chemical Protection, at No. 4a Samokatnaya Street (next door to the well-known vodka factory Kristall). This building was enclosed in a solid fence and surrounded by a park; the only decent approach was from the direction of the Yauza, a tributary of the Moscow River, but it was hard to pick out.

Shortly after the setting up of our new organizations—both the Interagency Science and Technology Council and Biopreparat—friction developed with the military side, whose members viewed civilian scientists with a jealous eye. The military not infrequently considered civilian scientists to be a bunch of dilettantes, and frequently argued with them during discussions of technological problems in the design and production of weapons. Besides, the military believed that they had more experience working with dangerous microorganisms—and, at first, that was really so. On the one hand, we had to prepare a scientific and technological "basis" for them, while at the same time Organization P.O. Box A-1968, the Fifteenth Directorate of the Soviet Army, was regarded as our "Customer." Its chief was Col. Gen. E. Smirnov till 1985, and after him it was Lt. Gen. V. Lebedinskij. The Fifteenth Directorate demanded the execution of all technical requirements, even if these proved to be unworkable at the preparatory stage. (I sometimes got the impression that these requirements were issued deliberately just to make us look incompetent.) For example, the military set us the general task of developing bacterial and virological strains with modified antigenic structures to overcome natural or vaccine-derived immunity. But the antigenic structure of even normal microorganisms had hardly been studied. There was little data at that time about human immunity to specific microorganisms, and in many respects the available data seemed contradictory. The military thought the problem might be solved with help from the Academy of Sciences, but Ovchinnikov, Skryabin, and Baev flatly refused to engage in the study of pathogenic microorganisms. This was most likely a good decision since they did not have the appropriate conditions in their laboratories for such studies.

In any event, every year the military issued a special order, setting out the chief fundamental and applied tasks for the coming year. The plans for these special orders were drafted by my own Department of the Council (i.e., Organization P.O. Box A-3092), but getting them agreed upon remained a struggle since every department kept itself under wraps. It called for enormous patience and diplomacy, neither of which I possessed. I therefore often sent my deputies in my stead, which was bound to be resented by my chiefs. The necessity for agreement among the various departments affected not only the drafts of major documents but also every decision by the Interagency Science and Technology Council; all such decisions had to be agreed to by all members of the council both before and after a session. In addition, these decisions also had to be approved "unofficially" by both the Military Industrial Commission and the Central Committee of the Communist Party, at which everyone jeered us. My colleagues from the Department of the Council (Organization P.O. Box A-3092) and I received no help from Biopreparat, whose members were not taken into the confidence of the Interagency Council. On their part, one heard nothing but mockeries, such as, "We'll see. Do you think you know best?" or "Let's see what you make of it!" I was utterly dependent upon this tedious and difficult process of obtaining official agreement. As a rule, as the head of the Special Department I could decide nothing; I could only offer suggestions. The last word belonged to various chiefs. Such was the Soviet practice of centralized authority. I was relatively independent only in my lab or when I was the director of some institute.

We had particularly serious problems over the allotment of foreign currency, some $10 million per year. The lion's share went to Biopreparat (40 to 45 percent), and just a little portion to the Academy of Science. In this sphere of gaining foreign currency the academicians were very active, which is more than one can say for their scientific results. Later, and largely out of these funds, Ovchinnikov raised a new luxurious building for the Shemyakin Institute of Bioorganic Chemistry, and Skryabin equipped his own institute at Pushchino-on-Oka. The remaining funds were allocated among the Ministry of Defense, the Ministry of Health, and the Ministry of Agriculture, with each receiving $1 to 1.5 million. The D. I. Ivanovsky Institute of Virology, Zhdanov's institute in Moscow, procured so much equipment that for a long time it was left lying about in the open. Altogether these authorities were trying to grab as much as possible without any fear of being accountable for procurements each year. This sort of overt graft was widespread in the Soviet Union. It did little to further scientific research, but a great deal to advance the personal goals and ambitions of Zhdanov, Ovchinnikov, and Skryabin.

Some of the money, from the Ministry of Health, officially designated for the plague-control institutes in Saratov, Volgograd, and Rostov, actually went toward "Problem Ferment." Had I been able to tell the future, I would hardly have agreed to the move to Moscow, since so many opportunities for work had opened in Rostov under the aegis of the new Problem "F" (even now I can only imagine what I could have done there).[15] When considering the foregoing account, however, one has to remember that the Soviet system obliged gifted and ambitious people to seek an outlet for their abilities. To do so they sometimes had to make compromises with their consciences and take jobs that would help to reveal their talents or enable them to achieve some sort of results in science. They did not feel obliged to worry about the means to their ends. Many people obtained what they desired in this way, including Ovchinnikov, Skryabin, and Zhdanov.

Of course, I could have found other work, but in Moscow it was very difficult because of obstacles set up by the authorities. Without their agreement neither Ovchinnikov (I mean the Academy of Sciences) nor the Ministry of Health could give me a decent job. I was too well known. Moreover, for most of my life I earned a very good salary, almost as much as a government minister, and had a big family by Russian standards. Once I missed the opportunity to move to Pushchino (near Moscow) at Skryabin's splendid new institute, but it was just after my appearance in Moscow in 1974, when I did not yet fully understand the nature of my work in the Glavmikrobioprom system.

Sessions of the Interagency Council were held in a "dead' soundproof hall to which the military people could come in civilian clothes. The military feared security breaches so the hall was checked before every session by the security service. One benefit of council membership was that each member received $500 annually for subscriptions to scientific literature. In those days journals and books were much cheaper than now, and I was able to subscribe to a lot of them. The secret service at Biopreparat examined all our journals, and gave permission for various subscriptions. When I moved to Obolensk I had to leave most of my books and journals behind. Moreover, I was only the one who could visit Moscow libraries! The secret service believed that by examining sets of journals and books one could discern the general direction of Biopreparat's work.

In 1975 the Interagency Council was reorganized. Evidently its existing composition seemed to the government not to be representative. Zhdanov was mysteriously removed from the chairmanship, indeed from the entire council, though not because of any defects in his work. He kept his post, however, as the director of the Institute of Virology. V. Belyaev became chairman of the council, while I was relegated to the rank of scientific secretary and head of the council department. Still, I retained my salary, personal car, and Kremlin telephone. To replace the men who had been removed, Smirnov and Ogarkov became members of the council.

It should be noted that the Interagency Scientific and Technology Council for Molecular Biology and Genetics was involved not only in the development of biological weapons for use against human beings, but also against agricultural animals and crops. Therefore, the Interagency Council set up a section for animals and crops: "Ecology" (Problem "E"), which was parallel to Problem "F" (or Ferment). I knew only that the Ministry of Agriculture intended to use different viruses for animals, and fungi as anticrop agents. A particularly secret department within the USSR Ministry of Agriculture that was similar to Biopreparat investigated the problem. The department served a wide network of research institutes that engaged in experimental, industrial, and testing work across the entire country. Just like Problem F, the Ecology Problem was an effort to produce genetically modified agents (germs and viruses) because the military believed that unmodified agents did not have a significant effect. Until the early 1970s, the Ministry of Agriculture worked with unmodified agents and the results left much to be desired. This special department's main virology institutes were located in Vladimir and

Pokrov, and the Phytopathology Institute was in Galitsino (near Moscow).

As we see, the stage was now set for the development of an ever expanding and more sophisticated biological weapons program built in the shadow cast by the Biological Weapons Convention. The United States had renounced its biological weapons in 1969, though we did not know this at the time, and even now I still find it difficult to believe. We saw ourselves engaged in patriotic work, advancing the study of molecular biology, immunology, and genetics in the Soviet Union, where these fields had been allowed to languish. Furthermore, working in these programs gave me an opportunity both to use my scientific abilities and inclinations to their fullest extent, and also to earn a decent living. Compromising with my conscience seemed, at the time, like a small price to pay.

NOTES

1. Regarding the epigraph: according to the *Shorter Soviet Encyclopedia* (1960, vol. 10, p. 1259), the aphorism of Vegetius "is often employed in capitalist countries in justification of the arms race and the preparation for aggressive war."

2. F. Fenner, D. A. Henderson, I. Arita, Z. Jezek, and I. D. Ladnyi, *Smallpox and Its Eradication* (Geneva: World Health Organization, 1988), pp. 366–67.

3. We still, in the year 2003, have a shortage of such basic materials!

4. See above, chapter 6, page 118.

5. The Soviet biological weapons program was huge and expensive, as can be seen by the distribution of corresponding institutions, including even new towns (e.g., Obolensk and Koltsovo). A considerable number of those institutions were built in the seventies.

6. Conference on the Prohibition of the Development, Production, and Stockpiling of Bacteriological (Biological) and Toxin Weapons and Their Destruction. Signed April 10, 1972, entered into force on March 26, 1975. The United States, United Kingdom, and the USSR were the cosigners and co-repositories for the treaty.

7. "Biological Weapons," *Scientist and Citizen* 9, no. 7 (August–September 1967): 153–67. According to Leitenberg's paper, the U.S. Army manuals list five groups of biological agents. We are concerned only with the first three.

Microorganisms: bacteria, viruses, ricketsiae, fungi, protozoa.

Toxins: microbial, animal, plant.

Vectors: insects, birds, other animals.

. . . What are the characteristics that recommended such choice?

- The agent must produce a disease that will incapacitate or kill.
- It should be capable of being produced economically in adequate quantities from available materials . . .
- Means must be available for maintaining the agent's virulence or infectivity during production, storage and transportation . . .
- Dissemination must be easy and effective . . .
- There must be no widespread natural or acquired immunity to the disease in the population to be attacked . . .
- Some forms of protection must be available to the user . . .
- Difficulty of detection adds to the military desirability of an agent, as does its ability to enter the body through more than one portal.
- There is always a lac or incubation period between exposure to an infectious organism and the first appearance of symptoms. All biological agents have some degree of persistence in the environment.
- Many biologic agents are contagious . . .

8. William C. Patrick, personal communication (Wendy Orent).

9. See page 118.

10. Klyucharev was an expert in rickettsia, one of the types of bacteria. Some of these bacteria are causal agents of spotted fever. The Military Institute at Zagorsk, where Klyucharev worked before his appointment to the Special Department, was engaged in the study of both viruses and rickettsia problems. The lengthy investigation of rickettsia was conducted under the direction of academician P. F. Zdrodovskij, who was at one time interned in a prison camp for "anti-Soviet activities." (See Glossary of Names.)

11. Moscow and St. Petersburg were always in competition with each other. To be an "Old Petersburger," i.e., Petersburg born and bred, is considered a sign of what we might call "noble origin" in the former Soviet Union.

12. See pp. 111–12.

13. It was impossible to discuss the morality of our work with my colleagues. Such a discussion was dangerous under the Soviet regime. At the very least, I could have lost my job.

14. The rhetorical question "What is to be done?" has considerable reso-
nance to a Russian audience. "Chto delat?" (*What Is to Be Done?*) is the title of
a famous treatise by nineteenth-Russian philosopher Nikolai R. Chernyshevsky.
V. I. Lenin borrowed that title for his treatise on the oppression of the workers
and peasants. (Ben Garrett)

15. Except for directors of the plague-control institutes, who would have
been told about Problem "F," however, all other workers officially knew only
about Problem No. 5. Their main task was the deep study of problems of virulence and the search for new methods of treatment.

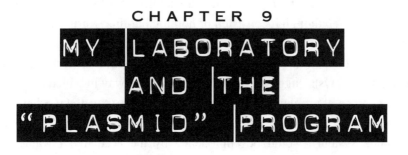

CHAPTER 9

MY LABORATORY AND THE "PLASMID" PROGRAM

Work in itself is a pleasure.

Manilius

I mmediately after I moved to Moscow in 1973, V. D. Belyaev gave me a small room in the Glavmikrobioprom building. None of the members of this establishment, with the probable exception of Belyaev's closest aides and the Secret Office,[1] knew who I was or what I was engaged in. I had nothing whatsoever to do at that time. According to Belyaev, work was supposed to begin after the issuance of a new order by the Party and the government. Until then I occupied myself by inventing some work. After many years of strenuous activity I couldn't just sit idle for days on end. The first thing I did was to write a letter, via the Secret Office of course, to the Minister of Defense, Marshal A. A. Grechko, which began with the words, "To Andrei Antonovich, with deepest respect." Next, referring to some documents available to me, I "earnestly" requested that he issue an annual report on the work done by the relevant sector on the program of biological warfare. On learning of this, Belyaev had a good laugh, since no one had ever ventured to address such a lofty individual in this manner, especially with requests to submit reports. But, to everybody's amazement, a report of a kind came back! True, it would have been difficult or even impossible to discover the truth of the matter from such a pro forma reply as the marshal had sent to me. Still, a reply is a reply, even though it consisted of nothing but fine words with no content at all!

I next tackled the Academy of Sciences and sent a letter to its president, M. Keldysh, suggesting that he examine the way in which the study of fundamental problems of immunology—that is to say, the mechanisms of both cellular and humoral immunity, the two basic arms of the human immune system—would be approached by the academy. I wanted to know what institutes would be set up to explore those problems. By this time Australian scientist F. M. Burnet had developed a clone selection theory that explained the nature of immunological specificity, or, in other words, how an organism distinguishes "self" from "foreign" antigens, particularly cells or proteins. Before Burnet, the cause of immunological specificity was unclear: how did the immune system of an organism distinguish its "own" cells from "what belongs to others"—i.e., invading pathogenic organisms? Immunological specificity is very important for the issue of organ transplants, and Burnet's theory has become a fundamental principle of immunology. Generally speaking, in this field of biology we were lagging behind the West by almost as much as in molecular biology and genetics: at that time (the early 1970s) we were mainly engaged in various improvements to the serological diagnostics of various bacteria and viruses[2] and we intended as well to study the immunochemistry or antigenic structure of bacteria essential for the development of new vaccines. But we were far behind in the study of the most fundamental aspects of immunology.

In the Soviet Union, the significance of Burnet's critically important theory was not grasped at all for a long time. We were behind in other ways as well, principally because we lacked the new information and understanding developing in the West. Our knowledge in this field was only on the level of what Western scientists had known in the forties and fifties. I was anxious to correct this problem and to bring our understanding of immunology up to that of experts in the West.

The reply to my letter was signed not by the president but by someone from two ranks below: A. A. Baev, who was academician-secretary of one of the divisions in the Soviet Academy of Sciences. He replied that "having considered the matter, the USSR Academy of Sciences regards immunology as a purely *medical* problem, and recommends that you approach the Academy of Medical Sciences."* This haughty attitude of Soviet academicians was typical, and exemplified the way they regarded the upstart, practical research of Glavmikrobioprom. To be sure, since our work was top secret, they could not have then appreciated the full significance of that research.

An institute of (clinical) immunology would nevertheless be created in another five or six years under the aegis of the Ministry of Health. Another immunological institute in the Biopreparat system was later formed, which I discuss below. The credit for the creation of these immunology institutes undoubtedly belongs to R. V. Petrov, who occupied the position of vice president of the Russian Academy of Sciences (RAS) after Ovchinnikov's death. As far as I know, the RAS has now ceased to regard the subject of immunology as a foreign body!

There had been very little current scientific literature available to me in Rostov and other outlying cities; now, in Moscow, I could spend all my spare time in libraries catching up on what I had missed before. I even began to write a book on the genetics of cholera vibrios, the motile, comma-shaped bacteria that are the causal agent of cholera. My deputy in the Department of the Council, Professor Klyucharev, likewise passed the time in libraries, having nothing better to do.

Just before my transfer to Moscow I had asked V. D. Belyaev about the chances of continuing my scientific work. He acceded to my request, and in the autumn of 1973 I organized a laboratory to investigate the extrachromosomal heredity of microbes—that is to say, to study *plasmids*, the strips of genetic material lying outside the main bacterial chromosome.[3] This was the first such laboratory in the country; it was located at Moscow's Research Institute for Protein Synthesis.

Why did this institute, where nobody had the least idea of either molecular biology or genetics, and not some other more suitable estab-

---

*I am quoting from memory.

lishment, become the base for my laboratory? It would have been more logical to set it up at the Genetics Research Institute, a leading institute of Glavmikrobioprom. But I never had that opportunity. Establishing my laboratory at the Genetics Research Institute was vigorously opposed by the prominent geneticist S. I. Alikhanyan, who had been on good terms with V. D. Belyaev since Glavmikrobioprom had been founded. Whether Alikhanyan was afraid of me (I was much younger than he) or whether there was some other reason for his enmity, I have never known, but from both him and his successor, Debabov, I met with intense opposition to my every move.

Belyaev may have acceded to Alikhanyan's insistence that my laboratory not be established at the Genetic Research Institute because he was squarely in Alikhanyan's debt. Alikhanyan had promised to produce many theoretical advances, which he would use to solve all scientific problems that Belyaev required. This inducement was very much in Belyaev's interest as head of Glavmikrobioprom.

Alikhanyan, as I well knew, could not deliver on those promises. I quarreled frequently with him and I considered him an unsavory character. I resented, in particular, his promises to solve any problem in the development of any microorganism as a biological weapon, pledges I knew he could not keep. I also resented how he played upon and manipulated Belyaev's trust. Alikhanyan had considerable influence over Belyaev, especially since he had already helped him in organizing the large-scale manufacture of yeast protein.

It is worth examining the personality of Alikhanyan, as he was one of Lysenko's[4] supposed opponents who attended the notorious session of the Lenin Academy of Agricultural Sciences (VASKhNIL) in 1948. (Lysenko had not yet consolidated his power until this session, at which it was still possible to criticize him.) Because of Alikhanyan's public stance as a defender of Mendelian genetics he was very popular with young scientists, who regarded him as a hero. During the new efflorescence of genetics in the Soviet Union, he took advantage of this popularity. Indeed, his speech at the 1948 session could have been called "fighting talk." He even conducted a diatribe against Lysenko himself. Hardly any of the young scientists knew the actual substance of this famous polemic, any more than they knew of Alikhanyan's criticism of the "erroneous" concepts of A. Serebrovsky, Yu. Filippchenko, N. Koltsov, and others concerning the gene. At this particular session the main argument had blown up about some individual principles in the Michurin doctrine, in particular on the occasion of vegetative hybridization.[5] "I take the liberty of acknowledging as fully justified the reproach that we are inadequately studying Michurin and Michurin's version of heredity and that we are paying too little attention to Michurin's methods," Alikhanyan said sarcastically at the meeting.

None of Alikhanyan's later adherents seemed to realize that only one day after his heroic outburst against Lysenko Alikhanyan retracted his own arguments and delivered a penitent speech. His words were typical for those days, "Beginning from this session, the first thing I have to do is to reexamine not only my attitude to the new Michurin science, but the whole of my previous scientific activity. I invite my comrades to do likewise." Not many other convinced opponents of Lysenko uttered a penitent speech! Alikhanyan obviously feared retribution and the loss of his post.

From another talk given at the VASKhNIL session by a certain Belenkij, I learned that as far back as 1939 Alikhanyan had promised in the press "to raise a new breed of chicken based on the Mendel-Morgan theory," i.e., artificial selection as understood in the West, but he never managed to do it. In his apology, Alikhanyan explained that, in trying to raise a new breed of chickens, he had been led astray by "objective circumstances"—that is to say, he had been blinded by the false promises of Western genetic theories. Now that he had "seen the light," of course, he backtracked from his promise of a "new breed of chicken" as expeditiously as possible.

It is evident to me, in retrospect, that Alikhanyan had always been full of bluster and knew how to play on the trusting nature of simple souls like V. D. Belyaev. My original assessment of Alikhanyan seems to have come pretty close to the truth. Still, in fairness to him, we should remember that in the days of the VASKhNIL session one had to sing "Hosanna!" to Lysenko and his minions with no reservations. Alikhanyan may have been obliged to "repent" in order to save himself and his colleagues.

In any event, I obtained my laboratory, which Belyaev felt should be used to tackle our problems on the nature of extrachromosomal inheritance in bacteria. The laboratory immediately attracted close attention from the secret service of Biopreparat, for obvious reasons: our work was of immediate interest and applicability to the research and design of biological weapons. In the study of molecular genetics I was, in effect, preparing my colleagues to solve a number of problems facing Biopreparat. Mainly we engaged in the techniques of gene transfer, techniques necessary for the development of the genetically modified strains needed to meet the requirements of the military. Biopreparat's secret service assigned a representative directly to me. As luck would have it, he turned out to be an old school friend of mine, L. A. Melnikov, who, being a microbiologist, helped in our work rather than just "looking on."

All of my reports went to Biopreparat, Organization P.O. Box A-1063. Special permission had to be obtained to release articles or papers for publication. Such publications were subject to additional scrutiny, which went well beyond normal censorship.[6] Publishing in Western journals was strictly prohibited. Because of this restriction I lost many opportunities to make the scientific community aware of my work.[7] This meant that I could publish only about ten articles in non-Soviet (Eastern Bloc) journals. These articles were carefully reviewed by the Biopreparat security service to exclude any possible leaks of secret information or even any allusion to my relationship to the development of biological weapons.

The laboratory itself was directly responsible to V. D. Belyaev, who had to approve all of my plans and reports. This was something of a formality, as he was not very well versed in scientific matters and relied entirely on me. His willingness to approve my work, without forcing me to submit it for additional approval by the heads of the Institute for Protein Synthesis, saved me endless trouble. Mikrobioprom, a technological institute with no connection to the problems of Biopreparat and biological weapons, ran the Institute for Protein Synthesis. So it would have been very difficult for me to explain to my colleagues at the institute exactly what I was doing: i.e., what kind of research my lab carried out, why I worked in that particular institute, and what was the source of my funding. Belyaev saved me from all such superfluous inquiries and conversations.

Although he was a product of the establishment, Belyaev had a generous and expansive nature. He was always pressing to enlarge the laboratory—at one point he sought to raise the staff to one hundred people—naively supposing, like a good "dialectician," that "quantity turns into quality." However, he finally agreed to fifty or sixty (all I needed). In 1978 I organized an interdepartmental genetics laboratory in Saratov[8] and later I organized a similar laboratory in the Krasnodar branch of the Institute for Protein Synthesis.

I also obtained some money from Belyaev for contract work. Unfortunately, most of it was lost through the slipshod attitude of the "contractors," although two contracts yielded fruit. First, there was the work done in the department of A. I. Korotyaev* at the Medical Institute in Krasnodar, who was able to develop his research on plasmids[9] and to obtain some interesting scientific results working with us over a fourteen-year period.

The second successful contractor for our laboratory was the genetics laboratory of A. Heinaru in Tartu (not far from the Estonian capital of Tallin), with which we carried out collaborative research over many years.

In my view, the most important of our results during this period (around 1970) was the discovery of the existence of plasmids in *Yersinia pestis,* the bacterium that causes plague, and the establishment of the role of these plasmids in plague virulence and infection.

Almost from the start of my work in Moscow, I began to hold meetings on the problems of molecular genetics. The first ones were held at Puschino with Skryabin, and were secret. Especially memorable was a

---

*Professor Korotyaev is a microbiologist. He took an interest in molecular biology, but had no funds. For several years I gave him money for the investigation of plasmids. He isolated a lot of very interesting plasmids of various pathogenic bacteria and studied their properties.

September 1974 conference at which some interesting people and noted scientists in the USSR at that time discussed at length the problems of gene transfer. Later, some of the attendees played greater or lesser roles in the development of this research (S. E. Bresler, A. S. Kriviskij, D. M. Goldfarb, A. A. Prozorov, T. I. Tikhonenko, V. N. Rybchin, R. B. Khesin, and others).[10]

Gene transfer is an especially significant series of techniques for solving certain intrinsic problems in the development of biological weapons, as well as for developing a general understanding of bacterial evolution. It involves transferring fragments of DNA coding for different properties of bacteria from one species to another. It may be achieved in particular through using plasmids, or phages (viruses of bacteria) to impart new genetic information to a bacterial colony. Technology also exists to integrate new genetic material directly into a bacterial chromosome. Gene transfer is used both in Russia and abroad for the development of new bacterial cultures or strains.

Fortunately, I was able to have all the proceedings recorded on audiotape, though the tapes were stamped "For Official Use Only." While I actually organized the conference, and even compiled the list of subjects for discussion, the establishment insisted that the meetings be held under the aegis of the Russian Academy of Sciences and presided over by Academician Baev. Once again my own position was highly ambiguous. On the one hand, I was one of the heads of a secret organization, and that was difficult to hide. On the other hand, I was a head of a well-appointed open laboratory and had the ability to organize conferences in different cities, an ability that was unusual then. Much of what I did in science was forced to remain secret even though I myself was plainly visible as a scientist. In 1974 the government adopted a new decree known as the Open Decree on Molecular Biology and Genetics.[11] As I discussed in chapter 8, a topsecret decree passed in 1973 established the Interagency Science and Technology Council on Molecular Biology and Genetics. This secret decree allowed for the pursuit of secret bioweapons research and design. Once the open decree was passed the following year, the secret decree acted under the cover of the open one. In other words, the "open" decree served as a convenient vehicle (a legend) to hide the true nature of the secret research. Y. Ovchinnikov[12] was appointed head of the Interdepartmental Council whose members controlled this "open" program.

I managed, through that open Interdepartmental Council, to secure confirmation for an All-Union "Plasmid" program. Plasmids are extensively used in genetic engineering to introduce new elements, such as antibiotic resistance, into bacteria—either virulent bacteria, for weapons purposes, or vaccine strains, for defense. Biopreparat supplied the funds for conferences and the publication of theses on this subject.[13] Despite objections from Alikhanyan and some others, the plasmid program endured for fourteen years. From 1978 on I held meetings under this program in different cities (Kiev, Tartu, Tallin, Krasnodar, Saratov, and Puschino), contriving each time to issue the papers by the beginning of each conference.[14] The 1982 Plasmid Conference in Tallin was even attended by a number of distinguished foreign scientists, including the American scholar Raymond Zilinskas, an expert in the study of biological weapons.[15] After this T. C. Gunsalos, a well-known biochemist and bacteriologist, invited me to attend an international symposium on molecular genetics in the United States as a co-chairman and speaker. The symposium organizers even arranged to pay for my visit. But the Soviet government did not allow me to attend. Once again, I had to "report sick" at the last moment.

Despite opposition from the Institute of Genetics and even from the Soviet Academy of Sciences, which had their own councils and special commissions, these conferences drew together many participants who also assisted in dissertation defenses.[16] Unfortunately, it became more difficult later on to hold these conferences, especially following Belyaev's death and after I left Biopreparat. With their funding source gone, the conferences stopped after 1987. Nevertheless I managed to arrange just one more in 1990 in Nalchik (in the north Caucasus Mountains), this time with the assistance of academician A. G. Skavronskaya of the USSR Academy of Medical Sciences, at the N. F. Gamaleya Institute of Epidemiology and Microbiology. This last conference represented a feeble attempt on my part to remain active in the field of genetics during the unsettled conditions in the Soviet Union at that time.

Sometimes I feel that it would be possible to trace the historical development of many aspects of the molecular genetics of microbes in the Soviet Union from the scientific papers published during those years. Indeed, if we had had control of the funds supposedly assigned to this program (which were actually held by Ovchinnikov) our work would have been far more effective. But there was never any question of directing the research, and we were left to analyze and generalize from what some enthusiastic individual scientists had managed to achieve on their own. These scientists worked far from the major scientific centers of research with minimal resources earmarked from the budgets of higher educational institutions or small research institutes. And to a large extent, these independent scientists were on the staffs of remote institutes. For them our conferences afforded the sole means of meeting and talking to one another about problems in molecular genetics.[17]

Under the Soviet system, funding for research was strictly controlled from the top down: before President Mikhail Gorbachev's "perestroika" policy, the government provided all funding. In the late eighties those funds were sharply reduced, and in the early nineties funding ceased for all practical purposes.[18]

Not all of my work was conducted under the Open Decree of 1974. In the next chapter I will discuss the beginnings of my secret work on the development of genetically modified strains suitable for military purposes.

NOTES

1. "The Secret Office" was engaged in receiving and sending off secret papers. These offices are found at almost every institute, even at universities, and in all state establishments, including departments. All secret information passes through them. This can also include business information.

2. One of the methods of diagnosis of infections and the identification of bacteria and viruses in the blood uses corresponding antibodies to see if the bacterium or virus binds to them. These tests are widely used in both Russia and the West.

3. The plague bacillus, for instance, contains several *plasmids,* usually

three, that carry important "virulence factors"—those genes which code for particular characteristics of the bacillus necessary for the development of infection in the host. These plasmids are strips of DNA located outside the main plague chromosome.

4. See pp. 61–63.

5. See chapter 2 for an extended explanation of Michurianism and Lysenkoism.

6. This scrutiny incidentally came under the jurisdiction of my department at P.O. Box A-1063.

7. I lost, for example, the possibility of contacts and exchanges with many colleagues, both within the Soviet Union and abroad; I could not attend a number of conferences even within the USSR, let alone in satellite countries or the West; I could not publish most of my work even in Soviet journals.

8. This laboratory was recently reorganized as a branch of the Genetics Research Institute (Moscow).

9. Plasmids are extrachromosomal elements of bacterial heredity. They are carriers of various genes, e.g., genes of destruction of many toxic compounds. They play an important part in the physiology of bacteria and are extensively used in genetic engineering to introduce new elements, such as antibiotic resistance, into bacteria such as plague.

10. See Glossary of Names.

11. This refers to the April 19, 1974, decree titled "Guidelines for Developing Molecular Biology and Molecular Genetics and Their Application to the National Economy. " (Ben Garrett)

12. See chapter 8, pp. 142, 145.

13. After I left in 1982 a special department of the Council on Molecular Biology and Genetics, Biopreparat at first diminished and then stopped payments. I received funds for the last time in 1987.

14. I failed to do so only once, in Saratov in 1983; the censor's daughter had failed her exams for the Medical Institute at which the conference was to be held!

15. At the present time, Dr. Zilinskas is a senior scientist in residence at the Monterey Institute of International Studies, Monterey, California.

16. As I have said, in those days publication was always problematic, but without publication one could not defend a thesis.

17. These programs were not "secret," nor were they directly related to bioweapons work. My "Plasmid" program was actually used by Biopreparat as a "legend" or a smoke screen for some of its secret warfare-related programs.

18. Naturally, the loss of funding affected my work. Since the fall of the Soviet Union I have not had the opportunity to conduct effective research. Those researchers, at present, who actually have such an opportunity have access to the ministries, the Academy of Sciences, and (especially) private firms. I lost my contacts with the Academy of Sciences and with the relevant ministries. The Academy of Medical Sciences, with which I am affiliated, ekes out a miserable existence. Many of our institutes are now in this impoverished and powerless state.

CHAPTER 10

THE WORK BEGINS

And in which I myself played a large part . . .

(Virgil)

I n about 1976, at my request, V. D. Belyaev applied to the Military Industrial Commission and the Central Committee of the Communist Party (the "powers") for permission for my staff and me to work on plague bacteria in the open institute where I now found myself (the Institute for Protein Biosynthesis). I had been working for three years on the study of extrachromosomal heredity (i.e., plasmids) in certain common intestinal bacteria, including *Escherichia coli*,* and *Pseudomonas*. Now I hoped to conduct similar research on plague itself. But this was not easy to do. My laboratory was in Moscow, and to work on dangerous pathogens— even avirulent strains of them—within the city limits was forbidden.

This prohibition was a direct response to the A. Berlin affair (see chapters 3 and 7), in which, as we have seen, several people died after an accidental laboratory exposure to plague. I wanted to conduct research using a particular avirulent strain, known as EV, which had been discov-

---

* *Escherichia coli* is the most studied bacterium, and a favorite object of molecular genetic research.

ered in Madagascar by two French researchers many years before. This strain retains immunogenicity (the ability to create immunity against plague), and is still used as a vaccine strain in Russia today. It is valuable for research because it is very similar in its physiology and antigen structure to virulent strains. This similarity makes EV a good model for virulent strains. Any genetic experiments we wished to perform on virulent strains, in pursuit of a more effective biological weapon, could therefore first be worked out with the much safer EV. These genetic modifications, the very basis for Problem Ferment, are known as "bioengineering." Such experimental manipulations on microorganisms, in particular on the plague microbe, were very attractive to the Soviet military. Therefore, there was much interest in my work. But how could I conduct such research in Moscow?

To my surprise permission for us to work with the EV plague strain at the Institute for Protein Biosynthesis was granted, and not just by anybody, but by the head of the KGB, Yuri Andropov himself![1] (I was even shown his actual decision: "Permission granted.") Obviously Belyaev had managed to pull off this feat somehow. There was, in truth, no connection between the dedicated purpose of the Institute for Protein Biosynthesis (which was in fact simple protein biosynthesis) and our proposed research with an avirulent strain of plague. Still, perhaps because the strain we wished to work on was indeed the avirulent EV strain we were granted permission. The KGB had ultimate control over all work related to Problem Ferment, particularly any research into the genetics of pathogenic (disease-causing) bacteria. The research, carried out in a populated area, was still theoretically dangerous, since it was impossible to predict the outcome of our experiments. Avirulent strains such as EV can, under certain conditions, recover their virulence and become dangerous again. But I did not believe such a reversion would happen, and I hoped for the best.

At some point after receiving permission, though I do not remember how it happened, I acquired a capsule of EV plague vaccine. In this form the bacteria have been freeze-dried (lyophilized), and they had to be brought back to life. This entailed growing the bacteria on nutrient media, including solid media, a procedure known as "performing a plating." With the assistance of my Biopreparat "guardian angel" L. A. Melnikov (purportedly the "senior scientist" at the laboratory), we isolated the culture, grew it on solid media, and the real work began.

This work involved, of course, one of the main aims of Problem Ferment, the development of genetically modified strains of plague that would be of potential interest to the military. My first task was to try to develop an EV strain with new and unusual properties—in particular, a strain capable of hemolysis (the destruction of red blood cells). This involved taking the genes from organisms that had this trait and implanting them into plague.[2]

The only trouble I encountered was when I had to inform some "cleared" members of the Institute for Protein Biosynthesis about my proposed work on the EV culture. Some members of my laboratory refused to take part in this new subject matter or, indeed, in any secret work. I could not understand the motive for their refusal: I could never discover whether they were afraid of something—perhaps the government's prohibition on travel abroad—or whether they did not wish to be involved in classified work. But in general they all came on board, and our first successes were not long in coming. In developing the hemolytic plague strain, we managed to transfer a cloned hemolytic plasmid (extrachromosomal genetic material) of an intestinal bacterium—*Escherichia coli*—into our colonies of EV. "Cloned plasmids" are artificially modified plasmids, sometimes with genes added to, or subtracted from, them. Among these added elements might be genes for properties such as resistance to antibiotics or different toxins, hemolysins, and so on. Our ability to use such a cloned plasmid to introduce a new property into plague was a significant achievement both on its own and as proof of a principle: the hemolysis plasmid, which causes the destruction of red blood cells, is not native to the plague bacillus. But it is an important virulence plasmid in other bacteria: our experiment therefore proved that alien genetic material could be introduced into the plague bacillus, something that had never been demonstrated before.

This experiment also had an even wider utility: the transfer of alien genetic material from one species of bacterium to another is critical in developing a military bioweapons program, because the successful transfers permit the creation of strains with new properties. This is the basis of all genetic engineering. The main tools, during this period in the mid-seventies, were various plasmids, including cloned, or artificial, ones. Using this approach, the military had the ability to create new strains of many different disease-causing bacteria. They could add new characteristics to the bacterium, including resistance to several different antibiotics or an altered antigenic structure that changed the way the bacterium or virus appears to the immune system. Altered antigenicity, in theory, allows the pathogen to evade an individual immunity to the disease, whether that immunity is natural or produced by vaccination. Strains with both altered antigenicity and antibiotic resistance would therefore make more people susceptible to a disease, and complicate the treatment of those who came down with it. The military applications for such altered bacteria are manifold.

Followed by a guard, I went off to see Belyaev to show him the dishes containing our new strain of genetically altered hemolytic plague. It was my first success in bioengineering.

At that time nobody knew what might be the outcome of experiments on the transference of alien genes even into vaccine strains. The general opinion was that it might lead to the emergence of bacteria with new, unusual, and perhaps dangerous properties, depending on which genes were introduced into the EV plague strain. We had to exercise the utmost caution, although in an open institute such as that of Protein Biosynthesis, which resembled a thoroughfare of common rooms, without any isolation cupboards, and with untrained personnel who had previously dealt only with intestinal bacteria, this was a difficult business. I can recall a situation that serves as an example of just how hard it was to operate under such conditions. It occurred to me to try to clone the genes of the purulence bacterium *Pseudomonas aeruginosa,* which code for the formation of hydrocyanic acid (a very powerful poison), and to transfer those genes to other bacteria without this property. I thought this property might be relevant to the Ferment Program. But to do so I needed to have the cyanides necessary for selecting the relevant recombinant clones.

Although I had no permission to work with the cyanides and couldn't get them officially, I consulted my friend L. A. Melnikov, and we decided I would try to produce my own hydrocyanic acid.

This was a dangerous project. Cyanide is a deadly poison, so we waited until we could be certain we would not be interrupted. After the staff had left the laboratory for the day, Melnikov and I locked ourselves in one of the rooms with a sliding cupboard and got down to work. However, in the course of our experiment the chemical reaction became too violent, and I didn't know how to stop it quickly. I finally succeeded by adding an alkali to the mix. But we were terrified!

There were other embarrassing and potentially dangerous situations. As I have said, working with domestically obtained toxins required special permission from the public epidemiological department as well as the police, both of whose conditions had to be met. Yet, at the same time (and somewhat paradoxically), we could freely import toxins from Western chemical companies, and nobody bothered about permits. Thus my laboratory accumulated a number of toxic substances that we were unable to use in current work. As soon as Dorogov, the party hack who was the head of the secret security services of Biopreparat,[3] got wind of it, he immediately ordered their destruction, although nobody could say how or where we should carry the task out. Dorogov was bold enough to suggest taking the poisons out of the city and burying them, but I managed to talk him out of that. So for many years these toxins lay in ordinary fireproof cupboards that could be opened with any number of keys.

This potentially dangerous situation could easily have been avoided had it not been for bureaucratic turf wars, the interdepartmental fragmentation and incompetence of our secret service, and a mania for secrecy, which was pursued to the point of absurdity.

In 1974, in conjunction with V. Zhdanov, the chair of the Interagency Science and Technology Council, I drafted a plan for what became known as the "Five Principal Directions." This effort to develop a new generation of biological weapons was approved by the entire Interagency Council in 1975 and confirmed by all relevant political and military bodies. The plan became known as "Bonfire."[4]

The object of this program was to direct and coordinate research at different institutions across the country on fundamental aspects of molecular genetics. Such research was necessary for the creation of new strains of bacteria and viruses. Our program also specified the safety requirements for research on genetic manipulation of these dangerous pathogens. The investigations at my laboratory were a part of this much larger program.

Our approach outlined the overall direction we hoped to take regarding the further development of strains of bacteria and viruses for Problem Ferment. In the past, before the establishment of the Interagency Council, biological weapons were based on natural strains of microorganisms. The ultimate purpose of Problem Ferment was to develop a second generation of biological weapons using genetically modified strains, which would be of greater military value. We planned to introduce new properties into disease organisms, such as antibiotic resistance, altered antigen structure, and enhanced stability in the aerosol form, making delivery of the agent easier and more effective.

We also reasoned that the genes for toxins or other factors could be introduced into the cells of bacteria or the DNA of viruses and thereby yield strains with wholly new and unexpected pathogenic properties. For instance, we thought of introducing toxins via plasmids directly into the plague bacillus. These plans eventually formed the basis for another of the major subprograms of Ferment, which would be called "Factor."[5]

But we quickly learned that producing pathogens with new properties is actually far more complicated than it seemed at first. Having acquired something new, a microbe would often lose other and more important specific characteristics such as virulence or stability. It must be remembered that microbes are living things, not mere strips of DNA where every coded property predictably emerges. Sometimes, when we added a new gene, we found that new, unexpected "emergent" properties appeared. This is because of the unpredictable genetic interactions between the newly inserted material and the entire bacterial genome.

I was often to come up against the same problem with other genetic experiments. In order to consider negative effects of these genetic interactions, we decided in the late 1970s to proceed in two directions; my own laboratory would continue to seek the means of transferring alien genetic information into the EV plague strain. Those institutes with the right safety conditions could directly engage in studying the genetics of virulence in the agents of highly dangerous infections, without which "Problem Ferment" could not be solved. The plague-control institutes and the N. Gamaleya Institute of Epidemiology and Microbiology had the requisite safety conditions, and they began to work directly with pathogenic bacteria in order to understand and clarify the genetics of virulence. This was an indispensable part of our overall program. However, since the scientists at these open institutes did not understand the true purpose of their investigations, and had to work according to the cover story they were given, their actual accomplishments in the field proved to be insignificant.

One of the first tasks of our redirected program was to produce a virulent strain of the plague agent that was resistant to a wide range of antibiotics.[6] The Research Institute of Epidemiology and Hygiene at Kirov, one of Smirnov's military institutes that had been working on plague, was selected to conduct the research. I was appointed head of the project (an unprecedented decision, since civilians were, as a rule, not admitted there).

I had to travel several times to the Kirov Institute, usually in the company of Maj. Gen. Igor Ashmarin (who was later responsible for the development of Program Factor) or Vladimir Lebedinskij, both of Organization P.O. Box A-1968, the Fifteenth Directorate of the Soviet army. Both men were deeply involved in the development of the overall Soviet biological weapons program. During my visit to the Kirov Institute I heard reports of the results of their work on the development of polyresistant strains and offered advice on the next steps necessary for their investigations.

The institute was situated in downtown Kirov and looked like an ordinary military unit. The personnel lived at the institute, which was carefully guarded. It also had a visitors' hostel. I managed to visit only a few laboratories, which were not outwardly different from those at the plague-control institutes. The military institutes had good state financing then: they received large additional grants through the council (about one million dollars per year). Therefore they were well equipped and had well-stocked libraries.

With their modern equipment, they were certainly better adapted for work on highly dangerous infections than my laboratory at the Institute for Protein Biosynthesis. I also remember the laboratory for aerosol vaccination, in which the procedures were accompanied by soft music (evidently to cheer things up!). I was greatly impressed by their library with its free access to the specialized literature and by the wide choice of recent foreign journals. This literature was never delivered to a scientist's home, because, under Establishment rules, it could be read only at the institute. People were also advised not to discuss their work at home. Therefore as the day's work ended at four or five o'clock, everyone engaged in his own personal activities (car repairing, fishing, hunting, or working on his summer house). The staff kept up the same lifestyle even after being placed on reserve duty, as my deputy at the Rostov Institute, who was a former officer from the Kirov Institute, told me.

It was no wonder, then, that, living in such seclusion, colleagues grew tired of one another. During one of my excursions to the summer residences of the institute, I was shown the house of one of the former directors, which had no windows whatever. This director, even in his dacha, wanted some respite from the need to look at his subordinates.

After work in winter I used to be taken to the theater, and in summer I would be driven into the woods or to swim in the Vyatka River. The then director of the institute, Maj. Gen. V. Pautov, likes to recall how he and I used to float down the river on logs. However, in those days, despite the idyllic surroundings, I was under constant surveillance. On one occasion, L. Klyucharev and I were traveling to Omutninsk (the location of one of the scientific and production bases of Biopreparat). This institute was not far from Kirov, and I decided to take the opportunity to ring the director of the institute from the post office. Our conversation contained nothing prejudicial, otherwise Klyucharev, a very cautious man, would have warned me. However, once in Moscow I was rebuked, since, as they pointed out, the mere fact of this conversation might indicate a connection between Biopreparat and the military. That the military and Biopreparat were actually connected was itself a closely held secret. Any connection between them, no matter how obscure, had to be kept hidden. Therefore, whenever I or anyone else had to go to Omutninsk, the military in Kirov never met us or accompanied us. Once in the dead of winter I was forced to spend the night in the train station because of these strict regulations, although the institute was only a few kilometers away, and the time until my Moscow train could have been spent more pleasantly.

In my laboratory the problem of devising a means of transferring alien genetic information—the transfer of genes from one species to another—to the plague microbe was soon solved. In our case this meant, of course, the transfer of genetic information from one closely related bacterium to another. In principle one can transfer any kind of genetic information into bacteria, including animal genes. In later years, this approach became the basis for the Program "Factor," which incorporated genes to produce mammalian or other animal peptides (small fragments of protein) in bacteria. This enabled bacteria or viruses to pump out large quantities of these peptides, such as interferons, interleukins, insulins, and small toxins, into their hosts, producing total disruption of the immune system and rapid death.

Some of the first steps toward this program were taken in my laboratory with the plague bacterium. Showing that foreign genes could be introduced into plague was a great achievement, since the plague bacterium has no means of its own for gene swapping. My colleagues and I had about ten patent claims to demonstrate our success. Unfortunately all of them are secret. I know what had been done and by whom, but it is impossible to recover the details from these papers. With no exact descriptions, I am not altogether certain whether anyone could ever make use of the patents as they are. If scientific papers are classified as top secret or secret, descriptions of their results are kept with the Secret Service; the authors merely receive certificates containing their names, patent numbers, and dates of registration. An author could read his papers at the Secret Service, but, should he leave his post, he lost this right. This eventually happened to me as well as to many other researchers.

The actual results of our experiments were reported to the Interagency Council and its various sections, where there were always some clever fellows from the Soviet Ministry of Health to communicate any incoming information to the plague-control institutes in Rostov and Saratov (these institutes had been allowed to participate in Problem Ferment, though in a very limited way). Consequently many of our results were soon reproduced at the plague control institutes, without any reference to my laboratory, my work, or the work of those who collaborated on the research. The secret regime of the military and Biopreparat was not bothered by matters of scientific priority (the data all belonged to the Establishment anyway, no matter who produced them). By now, of course, it would be impossible to prove who developed what. As a scientist this was a great personal loss to me. It was a loss for the country as well: international recognition for any major scientific achievement is valuable for the country where the discovery takes place. The ownership of scientific discoveries matters a great deal, since these discoveries may have great scientific and commercial significance.

But the greatest achievements of our laboratory came in 1977 with our discovery of the existence of plasmids in *Yersinia pestis,* the plague microbe, and in the closely related *Yersinia pseudotuberculosis*—and our demonstration of the central role of plasmids in plague virulence.[7] If plasmids are removed from the plague bacterium, the bacterium cannot infect and kill its host. Scientists in the West had long denied the existence of plague plasmids, and indeed they were not discovered in the West for another three years. That independent discovery propelled a rapid advance in the understanding of plague virulence among Western scientists, though whether it was ever used there for military purposes I have no idea. But the fate of our own, earlier discovery was the same as for the above-mentioned patent claims: our rights to claim scientific primacy sank into oblivion. Now, more than two decades later, all of this work could be declassified because the science has long since been superseded.

But in spite of my repeated efforts to have our work made public, the management of Biopreparat has never once answered me. Of course, to answer would mean that they knew about our work. No doubt that would cause quite an international scandal.

Among other achievements of our laboratory, which was not, however, directly connected with work on Problem Ferment, I would also include one of the first proofs of the possibility of transferring plasmids from a gram-negative intestinal microbe into gram-positive bacillus, as published in *Report AN USSR* (1976) and in the *Russian Journal of Microbiology* (1977). Initially, no one believed our reports, either here (the Genetics Research Institute) or abroad, although our discoveries were later confirmed by other people.

As regards the work done by Smirnov's department on obtaining poly-resistant strains, they did succeed so far as I know, although I was not given access to the final results. In their concluding report, which I signed, several strains with different spectrums of antibiotic-resistance genes were described, though no military evaluation of these strains was included. After that, our scientific relations with the military ended. I no longer undertook cooperative work with the Ministry of Defense. I never did learn what strains were chosen and used for subsequent work on biological weapons development. The reason for this abrupt cessation of contact was never explained to me, and I headed a program. I can only conjecture about the reasons for this concealment. I think it may have been due to the unwillingness of the military to share with me their secrets concerning the actual development of biological weapons. As I have mentioned, there was a long-standing competition between the Ministry of Defense and Biopreparat, and because of this competition there may have been a number of findings from which scientists like myself were excluded.

This marked the end of my direct alliance with the military. I was once again indirectly allied with them in the 1980s when I worked at the Obolensk Institute. They took part in the oversight of our work there, and made use of some of our Yersinia pseudotuberculosis strains, in particular one to which we had introduced genes from diphtheria (*Y. pseudotuberculosis* is closely related to the plague microbe).

In order to strengthen the laboratory in the Institute for Protein Biosynthesis, soon after work began on the EV plague vaccine strain, Belyaev suggested that I invite several people from the plague-control institutes to come work with me. At that time Biopreparat had a quota on Moscow-registered staff; under the Soviet system, no one could live in Moscow without specific permission and registration. For this purpose the laboratory was elevated to a department under the same title. It consisted of three laboratories. I invited four of my former colleagues (three from Rostov and one from Volgograd). But in 1979 Belyaev could get hold of only one three-room flat, which under the prevailing standards (meaning family size) fell to my student from Volgograd, L. A. Ryapis and his wife, who was also a researcher in highly dangerous infections. Ryapis moved in as head of one of my laboratories and I put his wife down for senior scientist at another laboratory, to avoid any "family collusion." As Ryapis was head of one laboratory his wife could not be employed anywhere at the institute; this would have been considered nepotism. I also engaged two of my colleagues from Rostov-on-Don who had earlier moved to Moscow. This helped to provide much-needed junior staffing. But I was unable to take on any more of my students: in the late autumn of 1979, Belyaev died, and everything changed, both within Glavmikrobioprom, and for me personally.

In the mid-1970s construction had begun on the new military research centers and institutes of Biopreparat, among them the Institute for Applied Microbiology (the Institute of Applied Microbiology) in Obolensk, a remote location about two hours south of Moscow by car. This was the only microbiology institute within Biopreparat devoted to bacteriology.* Research was conducted at a tremendous pace and in the deepest secrecy. Since after the 1975 ratification of the Biological and Toxin Weapons Convention it was illegal to research, design, and produce biological weapons, the true purpose of the laboratory at Obolensk had to

---

*There was another, parallel institute, the Scientific Research Center for Virology and Biotechnology, more commonly called "Vektor," which was established for the study of viruses. It was located in a town called Koltsovo, outside the city of Novosibirsk in western Siberia.

be carefully hidden. In order to mislead any foreign satellite intelligence the Institute of Applied Microbiology was deliberately built to look like some kind of sanatorium, with the convalescents walking about in pajamas or playing volleyball!

The Institute for Applied Microbiology paid great attention to the training of staff. They recruited graduates from the best colleges in the capital and other places within the Soviet Union (no foreign students were admitted, not even from the Warsaw Pact) and, until there was somewhere to work, they were sent away on temporary postings or given postgraduate status. In some cases, as payments for the training of our postgraduates, the administration of corresponding faculties and laboratories obtained equipment and reagents from Biopreparat. Most of these recruits were gradually called to work at the institute, although a few, and not necessarily the least proficient of those recruits, turned the assignment down on various pretexts and got themselves placed at the open institutes. There were different reasons for such refusals: some recruits did not wish to leave Moscow for such remote locations; others had personal reasons such as job preferences, marriage, a dislike of their accommodations, and so forth. Furthermore, some of them may simply have preferred to work in open institutes. They paid a price for this refusal: thereafter, they would be considered by the Secret Service as "untrustworthy elements." Such a label could prevent them from ever working in a closed (government-sponsored) institute.

This concern for excellence applied only to the junior and middle grades of researchers. Senior management positions were given to alumni of the Ministry of Defense system, mainly through personal acquaintance. This meant that corruption and cronyism easily crept into the selection process for heads of laboratories and departments. People suddenly appeared in these positions who were frankly not capable of much and who under ordinary circumstances would hardly have reached senior posts. But these were the people who supported the administration.

In 1978 the actual work on Program Ferment began at the Institute of Applied Microbiology at Obolensk and at our other as yet uncompleted institutes and centers. Belyaev appointed me as scientific director in

Obolensk. I therefore had to commute every week, which I continued to do for nearly four years. In order to work effectively, I transferred to Obolensk much of my personal library and a large number of altered strains that we had produced in the Moscow laboratory. I had an extensive collection of various bacterial strains necessary for genetic research. Many of these strains had been obtained from abroad. This collection was of considerable scientific value. Obolensk was a new institute and had no such library of strains. Gen. Yurin T. Kalinin, the head of Biopreparat, ordered the transfer of my collection to hasten the beginning of genetic research in Obolensk.

In addition, I brought on board several members of the Institute for Protein Biosynthesis to work with me at Obolensk. Every time I visited I assembled the laboratory heads—there were no "managers"—to discuss the work done over the week and to consider the progress of the experiments. We held these meetings in the laboratory of R. V. Borovik, a graduate of the veterinary institute in Kazan. Borovik turned out to be a very expeditious man who scrupulously attended to advice: he was a careful scientist. He was also politically astute in difficult situations, and got on well with administrative officials and party chiefs. Therefore, when he defended his doctoral thesis with only partial success, I used all my connections to prevent his failing before the VAK (the High Certification Commission).

My work at the newly created Institute of Applied Microbiology started off well enough. Had Belyaev lived, no doubt things would have continued in this positive and productive way. But it was not to be.

NOTES

1. Yuri Andropov was also the future general secretary of the Communist Party.

2. The introduction of novel properties into an organism has importance beyond that of proving an academic exercise. In this instance, Domaradskij writes of modifying the plague bacterium so that it has a novel property: destruction of red blood cells ("hemolysis"). That property could enhance the lethality

of the plague organism. Moreover, it could confound traditional or classical detection schemes, which might use the absence of hemolysis during laboratory culturing of an unknown organism as a property that distinguishes plague from other organisms. The net effect might be failure to detect modified plague bacteria when they are present. (Ben Garrett)

3. See chapter 8.

4. See K. Alibek, *Biohazard* (Dell, 2002), pp. 157–60.

5. See chapter 12, p. 206.

6. We always preserved in any agent sensitivity to one antibiotic, which could be used in case of a laboratory accident like that which befell A. Berlin.

7. Diploma for the discovery in *Plasma*, no. 001 (June 27, 1983) (priority established effective October 27, 1977).

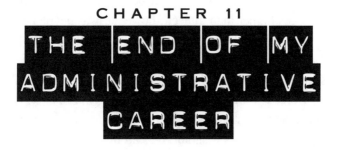

CHAPTER 11
THE END OF MY ADMINISTRATIVE CAREER

Don't bite off more than you can chew!

A fter the death of V. D. Belyaev, a bright and independent man, a great deal changed. His successor in the post of director of Glavmikrobioprom was the colorless R. S. Rychkov. Until his new appointment he had been sector head in the Defense Department of the Central Committee of the Communist Party, which used to oversee all of our work. Rychkov's "erudition" can be judged by his complete inability to communicate effectively with us until someone put together for him some special "glossaries" of terms and concepts. At that time immunology was a popular subject of research, which meant that I had to prepare a crib sheet with current concepts simply explained and neatly laid out so that Rychkov could understand what we were talking about. However, his incompetence never inhibited him from giving everybody his own opinion (even on immunology!), or from addressing his subordinates from on high in a casual, lordly manner. At the same time, with his chiefs and even with his equals he was pleasant and charming and was good at rounding off sharp corners and finding compromises. That may have been why he had risen to such a dignified position. In my view he was a poor administrator, and my colleagues may well have agreed—but of course this sort of thing could never be discussed. His ability to play up to supervisors higher than himself, however, was beyond dispute.

A new and fairly young man was also appointed head of P.O. Box A-1063, Gen. Yuri T. Kalinin, who till then had been director of the All-Union Institute for Biological Instrument and before that a member of the Ministry of Defense biological weapons institute in Zagorsk. Rumor had it that through his wife's family he had close ties with powerful people in high places. In all fairness it should be said that Kalinin is a very bright man with a good feel for political conditions, and he is well able to adjust and adapt quickly to new situations. It is these qualities that helped him to remain in place for years after the fall of the Soviet Union, and to deal more or less successfully with the post-Soviet transition from a planned economy to a free market.

Kalinin's deputy for science was Gen. A. A. Vorobev, also from Zagorsk, and a great friend of L. A. Klyucharev, the head of the Science Department of Biopreparat. Almost immediately after he arrived at Biopreparat, Vorobev instituted talks with the Academy of Medical Sciences. It seems that he had been promised membership in the academy upon his transfer to P.O. Box A-1063. As I describe in the appendix regarding "Academies," special places quite outside the normal procedures for election to the Academy of Sciences and the smaller, less prestigious Academy of Medical Sciences were reserved for outstanding members of the secret systems. Allotment of these special places depended on the rulings of the Central Committee of the Communist Party and the Military Industrial Commission. Shortly after the emergence of Problem Ferment this system was extended to us. Vorobev counted on this system for his election, since as a military man he was not known to the broader scientific community.

Vorobev was so keen on this academy appointment that he persuaded me to go with him to the Central Committee of the Communist Party to solicit their support in granting him such a special place in the academy.[1] The pretext for this visit was Rychkov's birthday, and Vorobev had prepared a gift for this occasion, some kind of a special clock. Rychkov accepted the gift with a promise of support for Vorobev's appointment to the Academy of Medical Sciences.

Relations between Vorobev and me began well, but they gradually soured; I resented his insatiable ambition and his boastful promises "to solve all of our complex problems in two or three years." In his arrogance he closely resembled S. Alikhanyan (see chapter 9). He seemed principally concerned with enhancing his reputation in his new place of work. I well remember how, at one of the council meetings where I was supposed to report on our activities in Obolensk, Vorobev announced that since he himself had been in charge of the science, things were going much better, and the outstanding task of producing an antigenically altered strain of the tularemia microbe would be completed in the very near future. He obviously wanted to be assertive and draw attention to himself, though the way he went about it was rather improper. He could not accept that on the subject of microbial genetics and biochemistry I was more knowledgeable than him. Vorobev's training and all his work at the Zagorsk Institute had been in virology with a little experience in immunology, but not in bacteriology. Nevertheless, he promised to solve what he considered to be the new problem of developing antigen-modified, and therefore "immune-resistant," strains of the tularemia microbe.

This was considered an important task: if a means could be found to alter the way a particular virus or bacterium appears to the human immune system, natural or vaccine-induced immunity could be overcome. Thus, the human organism might be made more sensitive to infections. It was certainly a principal aim of our work on tularemia, but Vorobev apparently did not realize that very little was known at that time about the normal antigen structure of the tularemia microbe. Producing an antigen-modified strain was well beyond his or anyone's capabilities at that time.

We also clashed over my proposal to try to isolate the gene for diphtheria toxin and to apply it to Problem Ferment. With complete assurance Vorobev insisted on the futility of my proposal, asserting that "adults don't catch diphtheria and it is not a problem in general." (The outbreak of diphtheria in Russia in the 1990s serves as excellent confirmation of Vorobev's incompetence, or of his habit of never shunning any means of getting his own way.)[2] I finally gained the upper hand by producing a novel strain of *Yersinia pseudotuberculosis* (a normally less pathogenic relative of the plague bacillus *Yersinia pestis*), which produced the deadly diphtheria toxin.

Afterward, in spite of our clear success, at any mention of my work Vorobev would merely shrug and never acknowledge my role. As the deputy for science at P.O. Box A-1063, he only needed to obtain permission from the public-health supervisory body for us to work on diphtheria at Obolensk, and also to obtain a toxigenic (toxin-producing) strain for us. Yet he seemed to take his time about it—for no reason I could understand—and I was forced to work "illegally," running the risk of getting into serious trouble. In order to work with any new bacterial agent we had to obtain the permission of the Third Main Directorate of the Ministry of Health.[3] To obtain this permission, I needed Vorobev's help, but he did not want to help me. Fortunately, the Third Main Directorate never learned anything of my work on the diphtheria toxin, or I might have received a reprimand, which could have resulted in a demotion or even my dismissal.

The details deserve some attention since they throw some light on the atmosphere of rivalry among the departments connected with Problem Ferment from the time of its inception. Having no means of obtaining the needed strains officially from the Soviet Ministry of Health (since Vorobev would not assist me in obtaining the strains or securing the requisite permission to acquire them) I procured, with the aid of my former wife, a specialist in childhood infection, some diphtheria smears from some of her patients in Saratov. At the Institute for Protein Biosynthesis (once again on the quiet) we managed to isolate some additional cultures, which I sent on to Obolensk. We worked with these cultures until we obtained a master strain that was both stable and reliably virulent. Despite our efforts, a scandal erupted at the Third Main Directorate and in establishment circles when V. Ogarkov* "unofficially" procured a strain of diphtheria and sent it to me at Obolensk. For some reason he had decided to help me, and he obtained the strain at the Moscow Central Control Institute, where there

---

* The first head of Biopreparat from its inception until 1979.

Sofiya Rotenberg (grandmother) and Ya. I. Drevitskij (grandfather) (Moscow, 1910).

Great-grandmother K. M. Drevitskaya (center) with my mother (right), Aunt Oksana (left), and my grandfather and grandmother (standing) (Taganrog, 1914).

"Counterrevolutionary gathering" in my grandfather's flat in Moscow (1927). All the family members are still alive. From evidence in the criminal trial (one of the "proofs" from KGB archives).

With my mother, Zinaida Domaradskaya (née Drevitskaya), the year I went to school (1933).

My father, V. V. Domaradskij (Saratov, 1936).

My grandfather a few days before his second arrest (Saratov, 1938).

Departure to Irkutsk (Saratov, 1957). Second from left is my future wife Svetlana Sergeevna Skvortsova; on the far right is my mother.

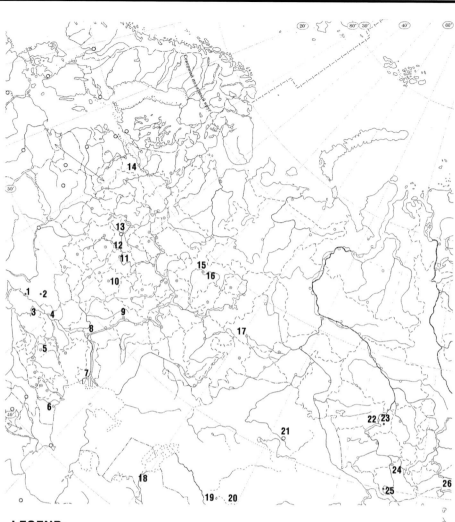

LEGEND:

1. Mariupol (in Ukraina)
2. Donetsk (in Ukraina)
3. Taganrog
4. Rostov-on-Don
5. Stavropol
6. Makhachkala
7. Astrakhan
8. Volgograd
9. Saratov
10. Tambov
11. Vladimir
12. Pokrov
13. Moscow
14. St. Petersburg (Leningrad)

15. Vyatka (Kirov)
16. Omutninsk
17. Ekaterinburg (Sverdlovsk)
18. Nukus (in Ucarakalpakiya, Uzbekistan)
19. Kzyl-Orda (in Kazakhstan)
20. Tashkent (capital of Uzbekistan)
21. Stepnogorsk (in Kazakhstan)
22. Novosibirsk
23. Koltsovo
24. Gorno-Altaisk
25. Kash-Agach
26. Kyzyl
27. Irkutsk
28. Kyakhta

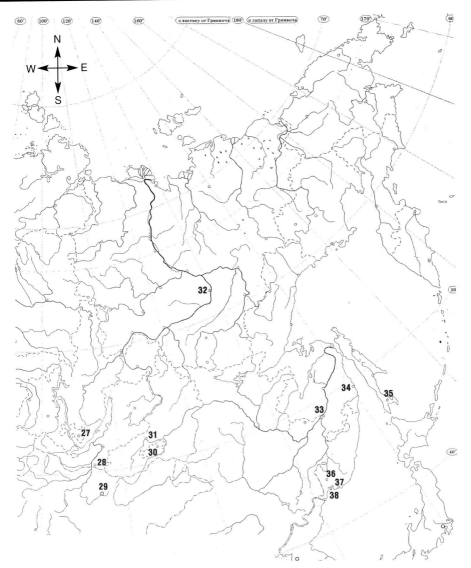

29. **Ulan-Bator (capital of Mongolia)**
30. **Borsya**
31. **Chita**
32. **Yakutsk**
33. **Khabarovsk**
34. **Vanino**
35. **Yuzhno-Sakhalinsk**
36. **Ussuryisk**
37. **Nakhodka**
38. **Vladivostok**

Notes:

Sergiev-Posad (Zagorsk) is about
60 kilometers north of Moscow

Obolensk is 120 kilometers south of Moscow

Lyubuchany is about 60 kilometers
south of Moscow

Galitsino is 40 kilometers west of Moscow

Me (on the far right) with (from left to right) Professor I. F. Zhofty, and Academicians of the Academy of Medical Sciences, USSR G. Rudnev, P. A. Vershnova, and V. D. Timakov at the entrance to the Institute of Plague Control (Irkutsk, 1957).

In China, at the Great Wall. From left to right, me, Dr. Di Shu-li, an interpreter, Prof. B. Fenyuk, Professor Petrov (1957).

Camp of experts in plague area not far from the Soviet/Mongolia frontier
(in marmota plague focus, 1959).

In Kyzyl (the geographical center of Asia). To the right are Prof. N. Y. Nekipelov and a
head of Tuva antiplague station, G. S. Letov.

In Mongolia, at one of the plague foci in the early 1960s. In the foreground are laboratory yurts and a truck of the antiplague service.

Mongolia, near its capital, Ulan Bator, 1960. At right is a head of the antiplague station in Kyakhta (in the Trans-Baikal area).

After hunting for capercailes (Irkutsk, 1962).

With V. V. Shunaev, one of the oldest members of the plague-control system (Irkutsk, 1962).

At a conference in the Chita antiplague station. At left is a head of the station, Galina Pletnikova, and on the right, Prof. V. Kraminskij (1962).

With the final-year students of the Army Medical College in Leningrad (standing) after their practical work on especially dangerous infections in the Rostov antiplague institute (1964). On my right is deputy director Maria Drotsevkina.

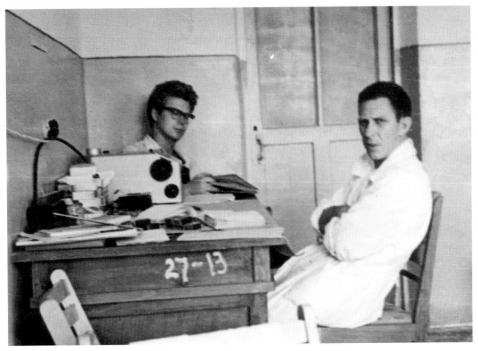

In the laboratory with my postgraduate student B. C. Rublev (Nukus, 1965).

After presentation of government awards (Rostov, 1965).
I was awarded the Mark of Honor.

Academician of the Academy of Medical Sciences N. N. Zhukov-Verezhmkov
(Rostov-on-Don, 1971).

Effluent drain channel in the center of Makhachkala (taken during the cholera epidemic
in 1970).

With my postgraduate students during cholera outbreak in Donetsk (Ukraina, 1971).

The construction of one of the laboratory blocks at Obolensk (early 1980s).

The first residential block on Biologists Avenue, Obolensk (early 1980s).

With my colleague Dr. Leonid Ryapis (at right) at the Institute of Applied Microbiology in Obolensk in 1982 or 1983.

Maj. Gen. N. N. Urakov, a head of the Institute of Applied Microbiology in Obolensk (1985 or 1986).

When the conversion began: Dr. Smolyaninov demonstrates technological equipment in the Institute of Applied Microbiology in Obolensk (the early 1990s).

was a large collection of different bacteria. On the way to Obolensk the culture leaked out. All that arrived at the Institute for Applied Microbiology was an empty test tube with a damp cotton-wool plug!

This misadventure, which did not really represent a significant health hazard, nevertheless backfired on a number of people, although the scandal did not touch Vorobev. Without special permission of the Third Main Directorate, we were not allowed to work with any new strains and we didn't have such permission. To obtain it, again, would have required the official request of Vorobev, who had already refused to help us.

In order to rescue Ogarkov from this embarrassing position, one of the professors dealing with diphtheria in Moscow issued a memorandum and sent it to the Third Main Directorate. This memorandum stated that Ogarkov was supposed to have received a "phagolysate" from the Moscow Central Control Institute, rather than the infectious strain itself. This was an extremely noble gesture since Ogarkov had previously received a reprimand from Leonid Brezhnev, then general secretary of the Central Committee of the Communist Party, for publishing what the regime considered a "seditious" book on *Aerogenic Infections* mentioned earlier in chapter 7.

This book revealed far too much about how infections are spread through the air or transmitted directly from person to person via the respiratory route. Aerosol transmission is, of course, the chief route of infection in a biological weapons attack. For revealing such information, Ogarkov's book was considered a "crime" by the Biopreparat secret service.

In spite of this memorandum, and mainly owing to "political intrigues" (I know that Ogarkov had many enemies, Belyaev in particular), Ogarkov was "relieved" of his work on Problem Ferment in 1982 and relegated to the position of simple deputy head of Glavmikrobioprom, where he was put to work on strictly civilian business. Since he had already succeeded in gaining some important contacts in Moscow, this demotion did him no harm: he was able to concentrate his efforts on acquiring full membership in the Academy of Medical Sciences of the USSR. This illustrates something of the schizophrenic nature of Soviet science: Ogarkov had a considerable reputation as a scientist in the "open" scientific community, and since Brezhnev's reprimand was top secret, it could not injure Oga-rkov's reputation among his peers outside the system.

During this period in the early 1980s, back in Moscow, Belyaev's death caused considerable upheaval, both in Glavmikrobioprom and in my life. Rychkov and I didn't get along from the start, and he soon found a way to force me out of Moscow.

The arrival of a new chief is inevitably accompanied by staff reorganizations. I have already said that I was not too keen on my job at the Department of the Council, P.O. Box A-3092, which was more administrative than scientific. I gave precedence to scientific matters, as I have always done. I realize now that in so doing, I was putting the bureaucratic management of Glavmikrobioprom in an awkward position and causing them considerable annoyance. Scientific issues were not their "subject matter"—although they always considered themselves infallible on any matter, even science—and Rychkov did not want to struggle with me over them.

Rychkov was nothing like the intelligent and generous Belyaev. Having risen from the lower ranks, Belyaev had learned to respect scientists and was happy about my work. Rychkov, on the other hand, had been reared in a different spirit. He was a Communist Party apparatchik, essentially ignorant about science. He was also very rigid, so we did not often understand one another.

Having decided to spare himself the necessity of daily dealings with me, Rychkov first proposed (in November 1981) that I undertake concurrently to direct the microbiological department in Obolensk (with most of my duties in the Department of the Council being transferred to my deputy). This was the first step in my removal from a position of central authority in Problem Ferment. A few months later (in February 1982), he ordered me to move to Obolensk and take up a permanent post as second deputy director for science. I was not at all pleased at this prospect. Moving to Obolensk meant that I would be placed more firmly under Vorobev's authority, though as a recognized figure I still retained more independence than Vorobev liked. He found my continued ability to associate with Rychkov and Kalinin exasperating, but there was nothing he could do about it, except for his petty attempts to make my life difficult.

To save me from any lingering doubt, Rychkov relieved me of membership in the Interagency Council. This transfer was managed very nicely: I retained my salary (through a special agreement obtained from the then chief of the Military Industrial Commission, L. V. Smirnov), my Moscow pass, and my department at the Institute of Protein Synthesis, as well as a flat in Protvino, and plenty of assurances on the real benefits I could bring to the common cause. I did not want to leave my work on Problem Ferment to which I had already devoted nine years of my life. Furthermore, to do so would have entailed significant penalties. I would face many obstacles from the authorities, who did not like those with secret knowledge leaving their control. No one—not Ovchinnikov at the Academy of Sciences or anyone at the Ministry of Health—could have given me a decent job without the permission of the authorities, and that permission would never have been forthcoming. I was far too noticeable, and had played a dominant role in the System for far too long.[4]

I had financial concerns as well. For most of my life I had earned a big salary by Soviet standards, almost as much as a government minister, and I had a big family. I wanted to continue providing them a comfortable way of life.

Apart from all those practical considerations, I found the science of Problem Ferment intriguing indeed. The attraction of that science seemed more important to me than what was to be done with its results. It must be remembered that, too, my work on biological weapons was carried out during the Cold War, and I saw myself as a patriot.

I had to accept Rychkov's offer. I arrived in Protvino (I refused to live in Obolensk itself) and lived there alone without my family for almost six years.

Given the central importance of science in my life, my removal from administrative work was probably inevitable. I consider scientific activity to be wholly incompatible with purely administrative work, which demands a completely different frame of mind and a special flair for getting on with the right people. I am not the only scientist to fall victim to this administrative/scientific tension. Another case in point was Prof. Yu. G. Suchkov, a true scientist, who was moved from his job as director of the plague-control institute in Stavropol in 1983 and sent to Moscow, where he was appointed deputy director of the Third Main Directorate in the Ministry of Health. After only a year and a half, Suchkov had to give up this job and return to scientific work. He just couldn't deal with the administrative obstacles.

In my own case, it was not a matter of "unsociability" as some have asserted. A man who has become used to thinking for himself and to finding his own way of solving his various problems is difficult to keep on a short lead. Therefore, whoever selected Burgasov over me as chief sanitary inspector of the USSR was quite correct to do so. You cannot fill two (or sometimes more!) chairs at the same time.

So now, though I found myself in Protvino, a remote community marginally more pleasant than the bleak Obolensk, I was now in a position to devote myself completely to science. This should have come as a welcome change, after all my dealings with administrative matters. But once again circumstances beyond my control quickly complicated my life, and I found myself in a position more difficult than anything I had yet encountered.

NOTES

1. See Appendix 1, "Academies."

2. Between the years 1992 and 1996 diphtheria had been contracted by 108,147 people in Russia!

3. See Appendix 2, "Institutes and Facilities."

4. In 1974 I had the opportunity to move to a distinguished laboratory at Puschino, near Moscow, which was run by Skryabin, but this was just after my first appearance in Moscow, when I did not fully understand the nature of my work at Glavmikrobioprom.

CHAPTER 12

"IN THE CITY
WITHOUT JEWS"

When the days of dreaming are past...

Pushkin

I n 1982 the director of the Institute for Applied Microbiology was Prof. Gen. V. D. Vinogradov-Volzhinskij, an "Old Petersburger." He was a medical doctor by training who had graduated from the Leningrad Academy of Military Medicine in the late forties. He became a parasitologist during a stint at the Zagorsk (now Sergiyev Posad) Military Institute in the fifties. Parasitology is very remote from molecular biology and genetics, but Vinogradov-Volzhinskij was a clever man with good administrative capabilities. He also knew his way about in Smirnov's domain in the Military Industrial Commission. In other words, he was well acquainted with the problems of biological weapons. He had left the army at one point, but later returned to active service when he moved to the Institute of Applied Microbiology at Obolensk.

Under Vinogradov-Volzhinskij's able direction the construction and organization of the institute at Obolensk went smoothly. As late as 1976, when I visited the site, I could see nothing but felled timber on the site of the future institute, while two years later there were huts on it (the "VLG" or the temporary laboratory village) in which people could work. The erection of the new buildings and residential quarters went on at the same time not far from the institute, which, at Vinogradov-Volzhinskij's suggestion, was named "Obolensk," after an ancient clan of princes who had lived in and owned much of the region. Here, the builders were army personnel, unlike the convict labor that built the virological research center at Koltosvo near Novosibirsk. I took no part in the design of the new laboratories in Obolensk and Koltsovo. This fell to the heads of Biopreparat and its secret service. But I knew that they wanted to create a laboratory complex that would not, in its external appearance, resemble the usual military laboratory in our nation or in other countries. They wanted to confuse any potential enemy who might be observing by satellite.

Despite his effectiveness as an administrator, Vinogradov-Volzhinskij was not well versed on matters of molecular biology and genetics, and he made no effort to become so. But in other matters which were the responsibility of the institute he held to one principle: "Small successes in every department, and then we will leave Efim [Smirnov's nickname behind his back] behind!" In other words, he felt that the Obolensk laboratory that he was directing could easily outshine the military laboratories. The two parallel systems were always in competition, and the Biopreparat wing of the Soviet bioweapons system always considered itself scientifically superior to the military division.

Since Vinogradov-Volzhinskij had no deputy at Obolensk and no knowledge of molecular genetics, Belyaev had appointed me in 1978 as his scientific consultant. Vinogradov-Volzhinskij was preoccupied at the time with building the institute and the residential sections of Obolensk. He therefore welcomed my weekly visits to the Institute of Applied Microbiology over a four-year period, and "farmed out" all scientific work to me. But he could not continue depending on a part-time deputy indefinitely, and when Rychkov came on the scene, Vinogradov-Volzhinskij's efforts to do everything on his own caused misunderstandings.

These difficulties forced Vinogradov-Volzhinskij to find a permanent deputy. At first he tried to solve the problem by appointing Prof. V. D. Savve, from Zagorsk, to the post. However, at about the same time, under the influence of Klyucharev, once my deputy at the Department of the

Council and now head of the scientific department at Biopreparat, a special institute of immunology was established at Obolensk with Savve as its director.[1] Vinogradov-Volzhinskij was, therefore, still without a permanent deputy. I suggested R. V. Borovik, who was one of only two doctors of science at the Institute of Applied Microbiology (K. Volkovoi was the other), and a young man I believed to have considerable personal initiative. Vinogradov-Volzhinskij thought this over for a few days—he scrupulously avoided any quick decisions—and eventually agreed. This happened shortly before my final transfer to the Institute of Applied Microbiology.

When Rychkov "suggested" that I should move to the institute, therefore, I was already thoroughly familiar with the circumstances there, both with the workers and with their research, and I expected to be given all the research facilities I had lacked since leaving the plague-control system. The institute had received modern equipment and all necessary supplies, including reagents, with help of the council. Most Soviet institutes did not have such equipment (except for a few institutes associated with the Academy of Sciences).

I had long since been cut off from the plague-control facilities. For reasons I am utterly unable to fathom, since my move to Moscow, the establishment had prohibited me from having any contact with the plague-control system. I was not even allowed to accompany my friend Prof. Werner Knapp, who came to the USSR at my earlier invitation on his tour around the plague-control establishment. I also could not send my book *Biochemistry and the Genetics of the Plague Agent* abroad, and colleagues at the Rostov Institute were even "advised" not to visit me at home once I had moved on to Biopreparat! This is because secrecy was a great concern at the three plague-control institutes, including Rostov, which were indirectly involved in work on Problem Ferment, as it was at Biopreparat-run facilities. Indeed, these plague-control institutes imposed a secrecy regimen not unlike Biopreparat's own.

Because of this mania for secrecy, therefore, I was able to visit the plague-control system only twice (in Saratov and Rostov) after my move to Moscow, with commissions headed by A. Baev. He made these visits for his own purposes, going to both institutes as if he were determined to control the results of their fundamental research in molecular biology. He introduced me as a representative of the Academy of Sciences, and not as someone from Biopreparat.

I think now, with the benefit of hindsight, that this prohibition on my visiting the plague-control facilities came not from those facilities, but from the Biopreparat secret service. They thought that my connection with the plague-control system might reveal the true purpose of the Biopreparat system, which was top secret (see chapter 8). They had the same motives for preventing me from publishing or even sending my published books abroad, and for blocking my connections with foreign colleagues. Our secret service believed that I would need to explain to those foreign colleagues what I, an expert in dangerous infections, was doing at Glavmikrobioprom in the first place. In my opinion this was completely absurd, but there was nothing I could do about it.

In the light of this peculiar prohibition the position of Col. K. Volkovoi seems paradoxical. Colonel Volkovoi was a rather conspicuous individual. Besides his scientific work, he was engaged in the selection of new recruits from Moscow and other university sites. He was familiar with many military bosses, because he had served at one of the military institutes. Before the Institute of Applied Microbiology, Volkovoi had worked at Biopreparat (P.O. Box A-1063). He had been transferred to the Institute of Applied Microbiology from Biopreparat for some serious infringement of the secret regime—I don't know what that infringement was—but he nevertheless enjoyed much greater freedom than I had. He was allowed complete "extraterritoriality," going anywhere he pleased, and never missing a single conference or a "school." I found this very strange. At the same time, he was engaged in recruiting staff for the institute. The selection of good younger staff can be directly credited to him. He was able to maintain his "contacts" outside the institute, which he used to find out what they were doing. If he discovered anything new through his scientific contacts or in his travels, he incorporated it into the work of his own department of general genetics and passed it off as his own achievement. His wanderings, therefore, often yielded new projects for his subordinates, and allowed Volkovoi himself to reproduce other people's experiments. How those who originated these ideas reacted to his tactics I do not know. Volkovoi did as he liked.

But oddly enough, and in contrast to many of his colleagues, Volkovoi was not a blatant careerist, and he didn't bother much about feathering his own nest. Like myself, he was living alone in Protvino, but he lived in a plain hostel rather than a flat. He did not appear to appropriate the ideas of others for his own direct benefit. Instead, he seemed more interested in filling in the gaps in his own thinking. Though he had started working at Biopreparat shortly after I did, he managed to keep his place longer than many high-ranking people, including those at the Ministry of Defense, with whom he was on familiar terms.

After my relocation to Obolensk, to my great disappointment, from the very start things failed to go as I had expected. First of all, Vinogradov-Volzhinskij refused to give me the laboratory (the one vacated since Borovik's appointment as the first deputy for science), to which I had become accustomed during years of visiting Obolensk, and where I already had some students. Second, and more important, I was allocated a very narrow field of work. General scientific management was under Borovik's control: this included matters that I undoubtedly knew better than he did, namely, biochemistry and genetics. I was to direct only one topic, the production of a genetically altered strain of the tularemia microbe. The director may have felt that I would be of greatest use in this area. This restriction to one topic irked me. I now understand that for Biopreparat it was very important to solve the difficult problem of developing an antibiotic-resistant strain of tularemia—and that they knew that I was the one who could do it. Obtaining strains of the tularemia agent that would be vaccine- and antibiotic-resistant for the subsequent development of a biological weapon was the first grave task of the Institute of Applied Microbiology. But I wanted to work on other problems as well, and I failed to see why I could not do so.

All the work of our establishment came under the public-health supervision of the Third Main Directorate of the Ministry of Health. Its representatives had their own views on highly dangerous infections and, although some people in the directorate had taken courses in Rostov while I was director, on a number of questions they paid no attention to me whatever. Owing to what they saw as the absence of proper conditions or trained personnel in the temporary laboratory village, for a long time they would not give us permission for any work on microbes, particularly altered microbes. If any work was to be done on pathogenic microorganisms, special safety conditions were needed. They were all the more necessary for work with especially dangerous bacteria, such as the agents of plague, brucellosis, melioidosis, and tularemia. Not to observe these conditions could be dangerous for the laboratory personnel and for the environment. Therefore, to research the tularemia microbe's biochemistry and its genetics, I needed new laboratories. At first these labs were housed in a small building, and then some of them were transferred to a large new building ("Building One").

Finally, in the early 1980s, when courses on highly dangerous infections had been organized at the Institute of Applied Microbiology on my initiative, the Ministry of Health granted permission for research on tularemia only, as a noncontagious infection. In general, the military preferred contagious infections such as smallpox and plague because they could cause epidemics with chain reactions in which one man infects another. Once introduced, they can spread widely without further effort on the part of the military. But tularemia is an effective agent as well, in many ways because it has a low infective dose, is very stable, and is easy to aerosolize. A strain of tularemia that could overcome vaccine immunity and be resistant to many antibiotics would make a formidable weapon. So it was felt that there were fewer risks involved with working with tularemia in the relatively ad hoc conditions under which we were forced to work.

This infection, therefore, immediately occupied a dominant position in the plans of P.O. Box A-1063. But very little was known about the tularemia microbe: we had no data on its biochemistry or its genetics. This lack of basic knowledge complicated our research protocols. We had to go back to square one and begin by creating two new laboratories: "Hut 7" for research on the tularemia microbe itself, and a "preparatory" lab headed by B. N. Sokov of the Ministry of Agriculture at their institute in Pokrov.

The Third Main Directorate of the Ministry of Health refused to allow any experiments with dangerous pathogens conducted by less than fully qualified personnel. So finding such qualified researchers was my first task. Fortunately, by the time I moved to Obolensk, one of the former staff at the All-Union Antiplague Institute "Mikrob," O. V. Dorozhko, who was very knowledgeable on dangerous infections (though not tularemia), was already there. So was my student Ryapis, whom I had previously persuaded to move to Obolensk. Having them immediately on board helped us get permission from the Third Main Directorate to proceed with our tularemia program. It must be remembered that, though Biopreparat was not a military program, it was nonetheless under the direct control of the Military Industrial Commission.

Furthermore, I obtained permission from the establishment to engage several of my colleagues from the Institute for Protein Synthesis to work on tularemia. These researchers were highly qualified in molecular genetics. This permit was not valid for long, and we got little benefit from the new staff (it was difficult for all of them to travel regularly from Moscow to the Institute of Applied Microbiology, two hours away), but one way or another we were able to begin large-scale research on our tularemia program.

Arguments about the reshuffling of spheres of command went on for months. Although Kalinin, when speaking privately, agreed with my desire to work on a wider scope than just research in tularemia, he either could not or would not try to persuade Vinogradov-Volzhinskij. Klyucharev, once my deputy, likewise made no attempt to help me take a more dominant role in the institute's scientific research, and was forever grumbling at my "unsociability" and "inability to get on with people." Everything suggested that he was influenced by Ogarkov and Vorobev. I have never been able to get along for any length of time with either man, and it is quite possible that they did not mind creating difficulties for me.

The military in general and the Military Industrial Commission in particular had put considerable pressure on us during the planning of our research. Biopreparat either could not or would not choose to object to that pressure. Sometimes, therefore, we had to incorporate some premature research designs that lacked any theoretical basis into our proposed plans. This resulted in some serious blunders in planning that soon became evident, and to a large extent Vorobev was to blame. He had made lots of promises he could not hope to keep and now he had to settle up. As always in such instances there was the inevitable search for scapegoats, and I was made one of them, though no one seemed to notice that after moving to the Institute of Applied Microbiology I had nothing whatever to do with the drafting of the five-year plans, which were the province of the Central Committee and the Council of Ministers.Uninvolved as I was in the actual overall planning process, I was a supervisor of those programs, and therefore a convenient scapegoat for Biopreparat, whose investigation exposed the unrealistic deadlines for some of our tasks, including the rapid production of a multiresistant strain of the tularemia microbe that could also "penetrate" a specific immunity, producing a vaccine-resistant strain.

These two tasks proved extraordinarily difficult, in part because we knew so little about the tularemia agent's genetic composition. Above all, our first essential task was to find a means to introduce alien genetic information into tularemia cells. After that, we wanted to use the same means to develop a multiresistant strain of tularemia, using antibiotic-resistant genes taken from other pathogens. And finally, we had to produce a strain of tularemia that not only was resistant to antibiotics, but could overcome vaccine-induced immunity as well. We were gradually approaching a solution to the problem of antibiotic resistance, but we were nowhere near the second goal. Nobody knew a thing about the mechanism of vaccine-induced immunity in tularemia or how the antigenicity (the surface components of germ cells or, in other words, how the agent appears to the human immune system) needed to be altered to overcome that protection. I still regard this as an unsolved problem.

The line of attack on this penetration of immunity—which had been planned as a result of an idea by Vorobev and the former wunderkind Zavyalov—was quite ridiculous. They had proposed to produce an alteration in the antigenicity, or outward "appearance" of the bacterium, by "stitching" the A protein of a staphylococcus to cells of the tularemia microbe! This amounted to trying to stick a piece of the staphyloccus germ directly onto the surface of the tularemia cells, which could not possibly have worked. Protein A would have no way to reproduce itself, since no genetic information for the production of the protein would have entered the tularemia cell. It's a little like sticking the wings of a crow onto a cat and hoping that the cat will produce flying kittens. Cells in the process of their reproduction would lose the new protein. On the other hand, if, instead of using the actual protein A itself, they had proposed synthesizing the corresponding gene and entering it into the chromosomal DNA of the tularemia microbe, the approach actually might have worked.

The assumption that we could develop a vaccine-resistant tularemia represented a significant error in our five-year plan. There was simply no way to overcome our fundamental lack of knowledge of tularemia's antigenicity. The task was therefore impossible, at least at that time.

These basic research problems, which prevented us from obtaining workable genetically modified strains to present to the military, threatened to disrupt their associated technological developments. I was out of my depth in that arena, not being a military type. What was more, after the failure of the five-year plan, I appeared to have grounds for insisting on a "reshuffling of the spheres of influence." However, in fighting for this I came up against one further obstacle. I was suddenly expected to work on an entirely new program.

At some point in the late 1970s one of Smirnov's deputies, Maj. Gen. Igor Ashmarin,[2] whose main interests lay in the field of physiology, had the idea of using genetic-engineering methods to obtain strains of bacteria that would produce various peptides, and in particular neuropeptides— short protein chains which, when expressed by a microbe, could produce particular physiological results. He suggested that these genetically engineered germs, without causing human death, might disable people, rather as biochemical agents such as tear gas, "intoxicants," or other "inoffensive" gases are supposed to do. It was a highly original idea, whose realization, at least in Ashmarin's original conception, could produce a more humane weapon than the deadly disease agents produced by Problem Ferment. This new problem, which became known as the "Factor," a division within Problem Ferment, had been approved by the authorities. The chief difficulty in its solution was not so much the need for a chemical synthesis or for DNA counterparts of the relevant peptides, but that we had to produce these engineered peptides in the cells of bacteria and in viruses and release them from the disease germs in a free form directly into the victim's bloodstream or tissues. In other words, bacteria and viruses would be used as pumps to produce mammalian hormones and other chemicals within an infected organism. Little is known even now about how genes of these peptides would be expressed or reveal their properties by the production of new chemicals, whether proteins, antigens, enzymes, or whatever.

The task of direct chemical synthesizing of peptides chemically was handed to the Leningrad Institute for Ultra-Pure Preparations (its public title) where the director was V. Pasechnik, who was to have a strong influence on the fate of P.O. Box A-1063.[3] The chemical synthesis of the appropriate strands of DNA—the actual genes that would produce these peptides—was handed to the Koltsovo Institute outside Novosibirsk because this facility was much more adept in molecular biology than the Institute of Applied Microbiology. The Koltsovo Institute was one of the first laboratories in the USSR to begin to synthesize polynucleotides (fragments of DNA and RNA). Those synthetic genes were to be used in Obolensk to obtain genetically modified plasmids. The synthetic plasmids were to be introduced into bacteria (e.g., *E. coli*) and then injected into laboratory animals in order to observe their effect.

This last part of the work was handed over to me, although my facilities were severely limited. We had neither physiologists to make the necessary animal behavior studies nor the expensive equipment necessary for physiological experiments. It was not long before these problems affected the whole course of my work on the peptide studies, and nothing useful emerged from this mess, so far as I was concerned.

☣ ☣ ☣

As for my living conditions, I was quite well off. In the center of Protvino, I obtained a splendid two-room flat on the ninth floor in an almost empty building, and I was allowed to use the "Ryabinka" grocery, which served only the directors of the High Energy Institute, the landlord of Protvino. In the early eighties, there were food shortages throughout the Soviet Union: people often had to go to Moscow for basic foodstuffs (e.g., meat, butter, and so on). In many areas something similar to the old wartime coupon system was introduced. Products such as butter, sausage, and meat were distributed using these special coupons. All that you could get on vouchers, at least in Protvino, was 300 grams of butter per month.

Nevertheless, these coupons did not solve the fundamental problem: there simply was not enough available food. In spite of these pervasive shortages, special grocery stores were opened for the elite, where weekly they could buy almost anything they wanted, including even red and black caviar!

For me the Ryabinka was redemption from starvation because I lived in Protvino without my family. The store even sold instant coffee! In addition, it delivered the products right to my door and took my next order. As the Communist Party so frequently proclaimed, "The age of advanced socialism" had already arrived! Shortages of all kinds continued until the collapse of the Soviet Union in December of 1991.

My family was living in Moscow. I was able to visit only on Saturdays and Sundays. This sometimes allowed me to "bag" Mondays, thus giving me time to visit the laboratory at the Institute for Protein Synthesis, which at that time had no need of daily supervision since my deputies were all highly experienced scientists. They needed only my weekly visits to maintain order.

At the Institute of Applied Microbiology the working day started at 8 A.M. In my youth I had been an "early bird," so that the need to get up just after five gave me no trouble. The day usually began with me putting on one of my favorite records and doing exercises, and in winter I sometimes went out skiing; luckily there was a park and a forest nearby. I also often went skiing after work. At 6 A.M. I nearly always switched on the radio to listen to the "voice from beyond"—Radio Free Europe, the BBC, or Deutsche Welle—which were not jammed in Protvino. Nobody bothered me, and I luxuriated in my freedom to listen to foreign radio, learning a great deal of news about the USSR and the world that was not available in Moscow.

In spring and summer I very much enjoyed car trips to the beaches on the Oka and to Tarusa or walks beside the river Protva. In this respect Protvino is well placed while Obolensk was much worse off. Gloomy forests and boggy ground surrounded Obolensk, although those in the know claim that the area has a lot of mushrooms and berries.

While living in Protvino I read a lot, and began to keep notes, which formed the basis of this book. I also did a lot of photocopying since I feared for the fate of my notes. We all were under permanent control of the secret service, and I could not take any chances: my not-always-loyal mood was well known to the service. I feared they would someday commandeer my notes, so I made photocopies so that I knew I would have a permanent record of my life and work.

My enforced solitude did not weigh upon me. In fact, it helped me to think. It was in Protvino that I began to think seriously about my way of living, and about the moral aspects of my work. I knew that I was part of a system that was out of control, but I could not think of an alternative way to live my life. Later, when circumstances forced me to leave Obolensk, I left that system, though in doing so I lost my status, and, eventually, all possibility of doing serious scientific work. Like my colleagues, I faced a terrible choice: remain with a corrupt and nonfunctioning system, doing work of (at best) a certain moral ambiguity, or sacrifice my entire scientific career. It took a long time and intolerable circumstances for me to make the latter choice.

The distance between Protvino and Obolensk was 16 kilometers. I was very fond of this road, which cut through mixed woodland. I liked it especially in winter when everything was snow-covered and you sometimes came face to face with elk.

L. Ryapis was living at the institute in the junior officers' hostel (they guarded the facility) and, like myself, he saw his family only on his days off. Other members of the institute lived here and there: in Protvino, in Serpukhov where we were then building some houses (it was 30 kilometers from the institute), in Pushino (60 kilometers away), and in the neighboring villages. My colleagues were brought by buses to work at Obolensk. The mass migration of scientists into Obolensk to live did not begin until much later in the mid-1980s.

Almost every month the Institute of Applied Microbiology was visited by a commission headed by Kalinin, with representatives from the Military Industrial Commission and the Central Committee of the Communist Party. They were also very often accompanied by V. Pasechnik, whom we have already met (in chapter 8). Visits by these directors always went off very ceremoniously. The institute had its own department of the GAI (the Road Police) through which all the other posts were notified, so that the directorate always had a "green light." They used to drive up with lights flashing and sirens wailing. The head of the institute's secret service usually went to meet them, joining the cortege at the junction with the old road to Simferopol.

The leading role at these conferences fell to me. The visitors listened to our results; discussed them; gave their "advice"; and, most important, reminded us of the need to finish the job on time. But that was easier said than done.

As I said earlier, we initially encountered some difficulty in our work, because so little was known about the tularemia microbe. That microbe does not recognize alien genetic information, nor does it possess the right genes for antibiotic resistance on its own. The problems inherent in the study of tularemia had discouraged geneticists before us. But several years of our own persistent work solved key questions of tularemia genetics. Having realized the cause of our previous failures, we eventually learned how to proceed. At the same time I well understood, as we all did, that our fundamental research into tularemia genetics would serve our "customer"—the military—in its quest to build biological weapons. But it was our work and we had to do it well. My scientific prestige depended on its results.

One of the small ironies of my life at Obolensk can still perhaps be seen in the city of Serpukhov. At that time my photograph was displayed on the Board of Honor in the center of town and my name was entered in the city Roll of Honor. If this Roll of Honor is still preserved, it will probably recount my achievements "in the development of vaccines and sera for combating dangerous infections and producing original means of plant protection." These lines of scientific research were proclaimed on enormous banners (faded with age) at the entrance to the administrative block of the Institute of Applied Microbiology. This was one of the clumsy attempts by the establishment to conceal the true direction of the institute's work from the public's gaze, because there had been no thought of developing vaccines or medical sera, at any rate not while I was there. There were actually some attempts to obtain new means of plant protection, but I had nothing to do with them.

Lest readers think that the Soviet Union was alone in its pursuit of secret biological weapons development, I would like to remind them that the United States, too, pursued such a policy for a very long time, until the program was shut down in 1969. No doubt they, too, had their own methods of maintaining secrecy.

At the same time, the plague-control system in the USSR, one of the most extensive in the world, continued the fight against dangerous infections. So nothing is as simple or so one-sided as it seems.

NOTES

1. At first, for a year or two, the Institute of Immunology was located at Obolensk, along with the Institute for Applied Microbiology; later it was moved to Chekhov, a town located between Obolensk and Moscow. Klyuchurev's friend Professor Savve continued to head the Institute of Immunology after its relocation.

2. Igor Petrovich Ashmarin (b. September 20, 1925, in Leningrad), like Viktor Zhdanov, has had a distinguished and varied career in the open scientific world as well as in the secret Soviet biological weapons system. His research interests include the biochemistry of contractile muscle proteins, the comparative biochemistry of the chromatin of various tissues, the determination of the phylogenetic links between bacterial chromosomal DNA and bacterial ecology, and the analysis of antibacterial and antiviral activity of chromatin. He also formulated a hypothesis concerning the evolutionary unity of various forms of biological memory (*Mysteries and Revelations of the Biochemistry of Memory* [in Russian] [Leningrad: Leningrad University Press, 1975). (Ben Garrett)

3. V. A. Pasechnik was one of the heads of the Leningrad Institute. He defected from the Soviet Union in 1989. His escape to England was a principal cause of the implosion of the Biopreparat system. See chapter 18.

CHAPTER 13
THE "AUTOCRAT"

There is no arguing with one who denies the principle.

From the Latin

A s I recounted in the previous chapter, despite some initial difficulties with the five-year plan, and despite my disappointment at being restricted to one field of research only, my associates and I were, by and large, content with our work. We were making progress at understanding the difficult genetics of the tularemia microbe. But everything suddenly took a change for the worse. Somehow Vinogradov-Volzhinskij had contrived to spoil his relations with Rychkov, and in September 1983 he was dismissed from his post as director of the Institute of Applied Microbiology. I think that Vinogradov-Volzhinskij's situation, and perhaps his personality in some ways, were similar to my own: Rychkov did not like independent and intractable people. Rychkov was a very hidebound man, and he and Vinogradov-Volzhinskij did not understand one another.

As a "consolation" Vinogradov-Volzhinskij was appointed to my old job as head of the Department of the Council (i.e., the Organization P.O. Box A-3092). I was a bit surprised at this move since I had not previously supposed that this post was at the level of general. Furthermore, the council had lost its former significance in the early eighties, and to

Rychkov it seemed less important than the directorship of Obolensk. The loss of Vinogradov-Volzhinskij was a setback for me. However difficult it may have been for me to work with Vinogradov-Volzhinskij, he was nevertheless intelligent.

About the same time G. V. Chuchkin died. He was the only man in the Military Industrial Commission who knew me since I had moved to Moscow and with whom I got on well. Chuchkin was one of the heads of the department of the Military Industrial Commission that managed all work on biological weapons, including Biopreparat. The council was under Chuchkin's control as well. We liked each other. He was also in sympathy with Vinogradov-Volzhinskij and he had been appointed one of the deputies to L. Smirnov, vice premier minister of the USSR, and head of the Military Industrial Commission. Chuchkin had possessed enormous weight, and not only in our sphere. After Belyaev, his was the second great loss for me.

On top of all this, my bacterial genetics department at the Institute for Protein Synthesis was gradually collapsing. After the death of Belyaev an unending shower of complaints began: people claimed that the department contributed nothing either to the institute or to Glavmikrobioprom. During Belyaev's time I had reported only to him and not to the director of the Institute for Protein Synthesis: in effect, my laboratory was wholly independent of the director and the institute. But soon after Belyaev's death and my move to Obolensk, I was informed that my reports had to be countersigned by the director, after being reviewed by the Institute of Genetics of Industrial Microorganisms. With one stroke, I lost my autonomy. Thereafter, I had to engage in some of the institute's research projects, none of which interested me.

Furthermore, the personnel in my department were gradually being laid off, and fear of job loss spoiled my relations with colleagues who no longer believed that I had the power to protect their positions. My repeated appeals to Rychkov and Kalinin fell on deaf ears, and only served to irritate the heads of the Institute for Protein Synthesis. The institute had its own reasons for wishing to commandeer my laboratory: thanks to Belyaev, I had modern equipment, and the institute wanted it, and my staff as well, to perform their own tasks. Had I stayed in Moscow I might have found a way out of this unpleasant situation. But as Obolensk began to manage on its own, my lab was no longer of interest to Rychkov and to Kalinin, and therefore they refused to intervene. Beyond lodging frequent complaints, I could do nothing but watch my laboratory disintegrate.

After Vinogradov-Volzhinskij's dismissal from Obolensk, Maj. Gen. N. N. Urakov was appointed director of the Institute of Applied Microbiology. There is no doubt that Urakov had more experience in biological weapons work than had Vinogradov-Volzhinskij. Urakov was a microbiologist and also understood the different aspects of bioweapons technology. Before this appointment he had served as one of the deputy heads of the Military Institute at Kirov. I had met him there when I was the head of a project to develop an antibiotic-resistant plague strain. He was a very thoughtful man who had listened attentively when we talked, and made notations on a big pad. So at first I thought that things at Obolensk would change for the better with his arrival. I hoped to be able to switch duties with Borovik and once again manage general science instead of being confined as I was to the narrow area of tularemia research. I wrote General Urakov several memoranda on the matter, but Urakov soon declared that my duties would remain unchanged. I would remain responsible only for tularemia research. He gave me no reasons. I was simply ordered to concentrate my research only on tularemia. He himself assumed direct control over the project.

Urakov soon became very involved in our work. Every week he would summon my colleagues and me and ask what had been done during the week and what results had been achieved. He always addressed me by first name while all the others were identified by their surnames, in the army manner. None of the meetings went off smoothly. Everyone went away cross and offended by Urakov's rudeness. It became immediately clear that we could expect nothing positive out of this change in command.

We had other grounds for conflict as well. My continued ties to a laboratory at the Institute for Protein Synthesis clearly irritated Urakov. Although the opportunity to visit this lab on Mondays had been a strict condition of my moving to Obolensk, he deliberately called meetings on those days, and each time he asked about my absence. He and his wife had settled in Protvino, and for a long time he was without a flat in Moscow. He would therefore call meetings on the tularemia project for Saturdays ("We are at work!" he would say), and I would never know in advance whether I could set off for home on Friday. Behind all this was Urakov's resentment of my relative independence from his authority. Furthermore, he had his own rigid views on the organization of scientific work: he felt that his directors and deputies should manage research projects rather than undertake the projects themselves. He did not want me to be directly involved in research.

For my part, I wanted my laboratory so I could continue to work on projects other than tularemia, and I wanted to maintain, as much as possible in so controlled a system, a measure of scientific autonomy and independence. Urakov bitterly resented this.

Being a soldier to the marrow of his bones, Urakov respected only force and brooked no arguments. His army comrades, several of whom worked in the institute, were accustomed to getting on with him, but for civilian scientists like me his rule was difficult to bear. For most of the institute's scientists there was no alternative: they were dependent on Obolensk for their flats and for those small privileges granted to them for their work on Problem Ferment. Urakov could not have evicted people from their flats, but he could deny new flats to those who did not have them. He could also deprive these scientists of new appointments or deny them the possibility of buying a car. (In Soviet times, procuring an automobile was difficult since hopeful buyers were put on a list for all available cars for sale. But Biopreparat's system had its own resources to circumvent the list. A member of the system could, if in good stead with his bosses, buy a car "out of turn" before his application to purchase one would normally have been authorized.)

For me and for my colleagues, however, the most difficult aspect of Urakov's regime was the complete neglect of fundamental science. Everyone who has ever dealt with the genetics of bacteria knows how complicated it is to produce a new strain, indeed, to create a new species!

In order to make Urakov realize this, we reported to him every detail of our work: how we obtained different variants and the methods we used. But he only said, "I don't need your strains, I need just one strain!" or "We are not playing here, we are making a weapon!" Then, at last, I realized the real purpose of our activities. Everyone in Obolensk spoke plain Russian: they had no qualms about using the most direct language to describe our objective. There had been none of this at the Kirov Institute! At this military institute scientific problems occupied the foreground while all questions relating to application remained unspoken. But now I understood that all of my knowledge was required merely to obtain a reliable and effective weapons strain of the agent, after which the "real" work would actually begin.

This stark openness at Obolensk may have been a shock to many of my colleagues, even though everybody appeared to accept the military's intentions. We all understood where we served and for whom we were working. As we Russians say, free cheese can be found only in a mousetrap. We knew that nobody, anywhere could receive the privileges we enjoyed without some strings attached. My colleagues and I knew full well what we were doing to earn our unusual advantages.

Urakov's arrogance was frequently intolerable. On one occasion, having listened to our report, he declared to me, "We produced the requisite strain of the plague microbe on our own while you, Mr. Corresponding Member of the Academy, can't do the same with the tularemia microbe!" But he was forgetting, first, that it was I who had helped "them" develop this plague strain, and second, that in terms of the knowledge of their biochemistry and genetics the two species couldn't even be compared.

When I was forced to transfer my whole collection of altered strains from my department at the Institute for Protein Synthesis to the Institute of Applied Microbiology as potential donors of the necessary genetic information (until then there had been no collection of live cultures at the institute), Urakov could not conceal his annoyance: "Why do we need all this rubbish?" he would say. He meant, of course, that my collection consisted of auxiliary strains, none of which in and of themselves met the requirements for being weapons grade. He wanted only a single, perfected strain. The rest he considered useless. This certainly showed Urakov's pragmatic, unscientific attitude: while these lower-grade strains, as well as the "intermediate" tularemia strains mentioned above, were "half-finished" products, they were far from being rubbish. In fact they were necessary for our continued work.

Urakov either could not or would not comprehend the essential problem facing anyone working with genetically altered bacteria: namely that, having acquired some new characteristic, microbes sometimes lose other characteristics that are equally important. We found this with the tularemia microbe. Having become resistant to several antibiotics, the strain lost its virulence, which was unacceptable to the military. They regarded even a drop in virulence from of one cell to two cells, or the protraction in the death of an animal by even a day as serious setbacks in the work. The desired bioweapons strain had to be fully virulent and deliverable in aerosol form. One germ cell had to be enough to start a lethal infection in a monkey. Furthermore, the infection had to be *incurable*. These goals were not at all easy to obtain. But I could not convince Urakov that the matter required serious effort and patience.

The matter ended with Urakov's decision to review the allocation of all jobs. He also invited new colleagues from the other military institutes to Obolensk to work with us. He believed we might all do better than my colleagues and I had done alone. He deprived me of access to almost all the laboratories except "Hut 7," the tularemia lab in the temporary laboratory village, and he removed L. Ryapis, who had directed Hut 7, claiming he was "not up to the job." Then Urakov handed me a former pet project of Vinogradov-Volzhinskij's, managing the group that developed the means of plant protection. The work on plant protection was one of the "legends" of the Institute of Applied Microbiology. Vinogradov-Volzhinskij had actually liked this work, and he knew the problem well. But for me, to be assigned to work on a "legend" instead of on my own research was a tremendous insult.

By then the Institute of Applied Microbiology had received permission to work on two other agents of highly dangerous infections, but these had already been passed on to other scientists. I was not allowed to "distance myself or divert from my existing assignment." In the beginning, the Third Main Directorate of the Ministry of Health permitted us to work only on tularemia, as a noncontagious infection. Eventually, after the facilities and the staffing had reached a proper level, we obtained permission to work on plague, anthrax, glanders, melioidosis, and modified agents. But this new work was not passed on to the original staff. Instead, Urakov invited his colleagues from the Military Institutes (as heads of corresponding labs) to work on most of these dangerous agents. He never completely trusted me or the other "civilians."

The results of such an attitude toward fundamental science were not long in emerging. Urakov's dogged and narrow pragmatism made him uninterested in other aspects of science, especially the basic molecular genetic research so critical to our work. He knew, and rated highly, only direct work on biological weapons. Urakov's actions were frequently detrimental to our research. Whether through obstinance or, perhaps, out of hostility to everything coming from me, he slowed the process of our scientific efforts. Very typical in this respect was his attitude to an idea of mine concerning the creation of "binary preparations." In the case of extreme genetic effects one often could not preserve all the desired properties of a culture, so together with my colleagues (including Kalinin, who was adopted in order to add "more weight" to the task), I suggested using two less "traumatized" strains whose overall properties met the requirements of the technical assignment. A traumatized strain is one that has undergone genetic manipulation and has lost virulence or acquired other undesirable properties. The more traumatized the strain, the less valuable it is as a biological weapons agent. I suggested, therefore, that we could have the same effect by combining two such traumatized strains that had suffered less genetic insult. The effect in the host of these combined strains would be a rapidly growing, extremely virulent, and essentially untreatable disease, which would bring about the same result as if we had managed to produce a single superstrain with its virulence and other properties intact. But even bringing Kalinin on board to assist—he was both coauthor of the binary concept and Urakov's boss—did me no good: Urakov was unmoved.

Even our "Customer," the Ministry of Defense, admitted the advantages of this method. But no one could convince Urakov, even though implementation of this idea would have helped to solve many problems and would have saved a lot of trouble, while the government's order would almost certainly have been completed. But out of sheer stubbornness, Urakov forbade this work to be done, so we were forced to go "underground." When asked the reasons for his negative attitude toward the idea of "binary preparations," Urakov merely replied that this was an old idea that had been known for ages, ever since the time of his (classified) doctoral thesis. But so far as I know, his thesis was devoted to the study of the combined effect of bacteria and viruses, which is a quite different problem. Interestingly, according to Ken Alibek,* Urakov eventually used my "binary" concept to produce plague that was resistant to about ten antibiotics. Of course Urakov would never have acknowledged that he had been wrong or that it had been my idea in the first place.

It is obvious, in retrospect, why Urakov seems to have come around to this point of view. The realization of our "binary" invention could make a fight with biological weapons more problematic because it would be hard to recognize its true antibiotic markers, and that might make treatment of corresponding infection very difficult. Besides, sometimes it is very difficult to create poly-resistant strains. My idea of using two strains, each made resistant to different antibiotics, could certainly help to solve that problem.

The fact was that one strain of any bacterium could carry too many added resistance genes, since it would inevitably lose other properties, mainly virulence. If a strain has four to six resistance genes, the problem is solved more easily. Each of the two strains can have three to four different genes and can be used together. In principle, in that case, it would be necessary for successful treatment of the patient to use between six and eight antibiotics instead of three or four. This would make countering a biological weapons attack all but impossible, especially on a large scale.

In addition to his rigid obstinacy, Urakov was utterly devoid of humor, as is shown in the following instance. The head of the staff depart-

---

*Personal communication with Wendy Orent.

ment in the Institute of Applied Microbiology at that time was a retired man, one Glukhov, who was totally hated for his Jesuitical character. One day there appeared on the notice board an order, signed by the head of P.O. Box A-1063, of a reprimand and withdrawal of a premium or raise in wages against Glukhov for vulgarity and incompetence. There was general rejoicing. At the same time there appeared an order from Urakov that everyone living on our premises in Serpukhov was to hand in his keys and apply for a flat in Obolensk. Naturally this produced a spate of phone calls to the director in protest (at that time living in Serpukhov was much easier than in Obolensk, and the communications with Moscow were better). But soon both papers proved to be an April Fool's joke. The director ordered them torn down immediately, and he called the very fact of their appearance a "political diversion." He started an investigation. One of the perpetrators turned out to be a postgraduate student of mine. Urakov refused to forgive him for a long time. In revenge, he almost spoiled the defense of the student's dissertation.

It may be difficult for the Western readers to understand such a situation. But during Soviet times, the position of chief of staff of a department was almost sacrosanct. He maintained secret files on all people who worked in the organization, and the destiny of each person was in his hands. One of Urakov's characteristic features was the total absence of a sense of humor or irony. That is why he could not admit that any jokes could possibly refer to him or to one of his subordinates. Urakov claimed the joke was a "political diversion" because that was the most awful sin imaginable in the Soviet view.

To provide a better overall picture of Urakov I should mention a few of his favorite expressions, such as "master the situation," "punish ruthlessly," "burn white-hot" (utterly destroy), "get realigned," and "the penalty is unavoidable." But if he were in a good mood, before signing papers he would ask, "Are there any faults?" This mode of discourse was entirely foreign to civilian scientists. For military people his specific phrases were normal and they understood each other very well. The military types would even repeat them in their own speech, because they all spoke one language. But for civilian colleagues these phrases were alien, and of course did not contribute much to effective departmental communication.

The words of Alexander Pushkin are very appropiate to Urakov, "In the capital he is a corporal, but in Chuguev* he is Nero." As we shall see, having Nero in charge of our premier bacteriological laboratory did not bode well for the future of the facility. Nor did it bode well for me.

---

*A small Ukrainian settlement.

CHAPTER 14

ROBBERY

Neither shame nor conscience...

French proverb

A t the end of 1984 another deputy for science, Col. V. S. Tarumov, appeared in Obolensk. Tarumov came from the Sverdlovsk Institute, a laboratory run by the Ministry of Defense and devoted to the production of weaponized bacterial strains, including anthrax. Neither Borovik nor I were specialists in the technological production of biological weapons. Tarumov, on the other hand, was an expert in this field, so he was extremely useful to Urakov. Tarumov was an active military man; he clearly thought highly of himself, and he claimed to have worked extensively with tularemia. I know nothing about the work he actually undertook with that agent, however; he never discussed it with me.

Tarumov was the same man who had been linked—how, exactly, I do not know—to the outbreak of a major epidemic of anthrax in Sverdlovsk, in which about seventy people died. In April and May of 1979, an unusual anthrax epidemic broke out in Sverdlovsk, a city of 1.2 million people located 1,400 kilometers east of Moscow. Soviet officials immediately attributed the outbreak to the consumption of contaminated meat. U.S. agencies attributed it to inhalation of spores accidentally released at a military microbiology facility in the city. Epidemiological data show that most victims worked or lived in a narrow zone extending from the military facility to the southern city limit. Farther south, beyond the city limits, livestock died of anthrax along the length of the zone. In 1992 the Yeltsin government permitted a special commission of American scientists headed by Harvard biologist Matthew Meselson to come to Sverdlovsk to study the outbreak. They concluded that the escape of an aerosol plume of pathogenic anthrax from the military facility caused the outbreak.[1] Apparently there was an explosion at a production facility at the laboratory; a technician seems to have forgotten to change a filter on an armored aerosol chamber and a small amount of anthrax, perhaps less than an ounce, blew into the air.

Despite Meselson's findings, the Russian military even now officially denies any hint that it was implicated in the tragedy, although in private conversation many of them insist on the opposite. Not wishing to denounce anyone, I will not name those who arrived in Obolensk and recounted stories about how, during the outbreak in Sverdlovsk, military personnel had been going to the houses of the victims disguised as doctors and withdrawing death certificates, allegedly in order to "refine the diagnosis." In 1979 the Soviet government had managed to keep the epidemic quiet: almost no one in the country, apart from military doctors and civil medical staff in Sverdlovsk itself, knew anything about it. This was the general practice: the government always tried to keep news of any outbreak from its own people. But, ironically, the rest of the world learned about the outbreak rapidly: reports leaked within a few days to the outside world.

Strange as it may seem, Gen. P. N. Burgasov himself, who was in charge of the 1979 Soviet investigation of the anthrax outbreak, to this day denies the role of the military institute at Sverdlovsk in the incident. Burgasov, who remains a true Soviet functionary even now, continues to insist on the disease having been imported! He maintains the "meat" explanation even now, claiming that contaminated meat had been imported to Sverdlovsk from some other region of the Soviet Union, thus causing the outbreak. Why the pretense? Is it a question of loyalty to the military, or an attachment to the ideas of a collapsed system in which he was brought up? Or it is possibly reluctance to admit in public that deliberate deception was carried out and that some people were forced to lie?

To this day, the Russian military offers two versions of the outbreak. Sometimes, despite the fact that the victims died of inhalation anthrax, they still resort, as Burgasov does, to the totally discredited notion of "contaminated meat"; at other times they claim that the entire tragic misadventure was "an American diversion" to embarrass the Soviet government.

Even before Tarumov arrived at the Institute of Applied Microbiology, another "specialist" was transferred to Obolensk from Svedlovsk. This was one Gennadi Anisimov, Tarumov's subordinate, who was introduced to us by Urakov as a leading specialist in tularemia. Tarumov claimed that under his leadership at the Sverdlovsk Institute, Anisimov had created a tularemia strain that met the military's requirements. Since my work on obtaining such a strain was not yet finished, both Urakov and Tarumov hoped to bypass my research altogether and proceed immediately to begin work on the development of a biological weapon based upon that strain. They hoped in this fashion both to enhance their own prestige and to show my inability as a civilian to achieve military objectives. This was another example of the long-standing competition between Biopreparat and the military laboratories run by the Ministry of Defense. It made no difference to the Military Industrial Commission which institute or laboratory solved the problem, so long as it was in fact solved. Perhaps, indeed, they encouraged this competition as a way to insure against any failure.

Many meetings were held at which regular arguments broke out about my tularemia strains. At one of these, Tarumov suddenly announced that he could see no point in my work, because the desired weapons-grade strain had already been developed in Sverdlovsk, and we had only to ask for it to be sent over. This statement puzzled us. If it was true, then why did we have to work so hard, wasting so much money and effort, just to duplicate the Sverdlovsk results? Urakov thereupon announced his "will": let all further work on the production of the enhanced tularemia strain be passed over to Tarumov and Anisimov since they were the more experienced specialists who could quickly reproduce the Sverdlovsk results. The tularemia department was soon moved to a new block and put under Anisimov's control. I was left with only my close assistants. I should note that this decision won no support in P.O. Box A-1063, including its director, Kalinin, but it went no further than their expressing some sympathy with me. Nobody wanted to argue with Urakov about it. Urakov did not mind taking such an action without the support of P.O. Box A-1063. Evidently he did not expect any repercussions: he seems to have thought, "a victor is outside jurisdiction." If he succeeded in his objective no one would dare to criticize him.

At the end of 1984, by means of various contrivances, my colleagues and I had finally managed to obtain two tularemia strains with the right properties. We found a way to produce antibiotic-resistant tularemia strains that retained their virulence, but the spectrum of their drug resistance was rather narrow. Furthermore, each of these strains had certain defects from the military's point of view, and each needed technological trials. Kalinin announced, after our report at one of the meetings, that "the task had been completed, though not to the full extent." It was therefore decided to release one of the strains for further work, and then to proceed with testing the corresponding preparation for its utility in weapons development, although this was now a field in which only Urakov and Tarumov were genuinely well versed. The tests went ahead and some success was achieved. However, some new defects in the test strain arose that I could not have foreseen. The tests on monkeys did not go as well as I had hoped: when they were injected with the test strain, their resistance to tetracycline was not high enough and they died too fast. In addition, there were some technological difficulties with the test strain, including poor growth on nutrient media, among other problems. Furthermore, Anisimov and others who had Urakov's confidence conducted some of the tests. I did not trust their results, however. In any event, with sufficient time, I could have anticipated and corrected every possible defect that did arise—but I was not given that time. Tarumov claimed that his strain was already free of any such imperfections. This gave Urakov the excuse he needed for a redistribution of responsibilities.

A month after Anisimov's appointment as head of the tularemia department I happened to learn that he had achieved a high resistance to tetracycline, the "stumbling block" of all our work. In our strains, resistance to tetracycline had never been high, and all attempts to raise it resulted in the decrease of virulence. At the same time, tetracycline is an antibiotic with a broad antibacterial spectrum; it is therefore widely used, in particular by the military, for the treatment of various infections. So the production of a tetracycline-resistant tularemia strain was a matter of considerable importance, which had nevertheless eluded us to that point.

Naturally Anisimov's announcement came as a surprise to my colleagues, and rumors circulated that in fact he had brought this strain with him "in his pocket" from his previous workplace, though Anisimov claimed that he had done it all "from memory."

While on a routine visit to Moscow I reported this success by Anisimov to Vladimir Lebedinskij, who had just replaced E. Smirnov as head of P.O. Box A-1968 (of the Fifteenth Directorate of the Ministry of Defense). Lebedinskij's opinion was unambiguous, "He is a robber. He pinched that strain from us!" (Lebedinskij meant that the strain had been taken from the Sverdlovsk Institute, where Anisimov had worked before his arrival at Obolensk.) The reaction was instantaneous: the next day a commission from Moscow rolled up at the Institute of Applied Microbiology to investigate the circumstances of this "theft."

Pursuant to the then current "Regulations for working with the agents of highly dangerous infections," in which the agent of tularemia was classed, it could be worked on only in special establishments, from which it was categorically forbidden to remove any cultures. The theft of a strain from a military institute could be interpreted in no other way than as an "emergency situation"!

But the investigation dragged on because Anisimov continued to insist that "recreating the strain was easy," and he was obviously backed up by Urakov and Tarumov. He repeated his claim that he had not pinched the strain from Sverdlovsk, but had actually recreated it from memory at Obolensk. A commission of the institute was set up to compare my own strains with those of Anisimov. In the course of deliberations it was discovered that, unlike our strains, Anisimov's had one important mark that was peculiar to the Sverdlovsk strain, namely, sensitivity to nalidixic acid. (It was also resistant to other preparations normally used to treat tularemia.) Nalidixic acid is one of a class of antibacterial chemotherapeutic medicines. It may be used for the treatment of several infections, including tularemia. If a strain was sensitive to nalidixic acid, it might be employed to treat USSR soldiers caught in their own tularemia bioweapons outbreak, when all the usual means for countering the bacteria—the measures used by one's enemies—would fail. It could also be used for treatment in the event of a laboratory accident. Our strain had sensitivity to some "reserve" antibiotics as well but not to nalidixic acid.

When Lebedinskij learned about this sensitivity to nalidixic acid, he sent another commission (in January 1985), but they didn't find Anisimov's strain. By then it had been eliminated from Obolensk's strain inventory. It had been "killed" by Anisimov to prevent any further investigation. Nevertheless, the theft of the strain from the Sverdlovsk Institute could be considered proved.

But events didn't work out as one might think. I don't know which member of the commission was responsible, but in order for the military to save face it was officially announced that the strain had actually been stolen, not from the Sverdlovsk Institute, but from my own team! It was stated that Anisimov, taking advantage of his position, had obtained the strain from the inventory of live cultures in my department and passed it off as his own. It seems that Urakov and his friends from the Ministry of Defense concocted this explanation, which implied that Anisimov was guilty of thoughtless behavior and the abuse of his position at the Institute of Applied Microbiology rather than of the actual outright theft of military property. They must have feared that without such a story the truth of the matter might have become known to the Central Committee of the Communist Party and the Military Industrial Commission.

In spite of all this, Anisimov came out of the ordeal relatively unscathed. Urakov merely punished Anisimov with a severe reprimand for conduct "unbecoming" to a scientist. The order was read out at a session of the Academic Council and the theft received wide publicity.

However, some military face had indeed been lost, to which Urakov could never be reconciled. He decided to work off his resentment and embarrassment on me. At a meeting of the Academic Council soon afterward, he referred to the results of the commission's inspection of our strains and to the possibility of their future use. It was then suddenly announced that the two strains did not demonstrate the characteristics we had claimed for them, and that the authors "had deliberately attempted to mislead the management of the Institute of Applied Microbiology and P.O. Box A-1063"! Urakov claimed that the commission, whose members had inspected my strains, had uncovered several defects, which he said I had tried to conceal. It was nonsense. But Urakov put on a good performance in front of the Academic Council. It was enough to discredit me.

Then Urakov asked, "Gentlemen, what are your proposals on the matter?" He then stood aside so as "not to exert any pressure on his colleagues." The Academic Council was composed of the heads of laboratories and departments who were totally devoted to the director. It was no wonder, then, that all the "colleagues" who spoke proceeded to advise him "to penalize me severely" (including a reprimand in terms of the party line), although some of them were people who during the early years of the institute had studied the principles of microbiology and genetics under me. Even the vivarium-keeper spoke up, admitting first that "though not yet a scientist, he was training to be one," but that nevertheless he had long suspected me of trying "to hide the truth," and he therefore sided with the rest.[2]

In his concluding remarks Urakov expressed his satisfaction at the "mature level of consciousness in members of the Academic Council, in being able at short notice to deal with everything so promptly and well." However, he recommended that they be "less severe" and proposed that they confine themselves to subjecting me to "public reprimand."

Understandably, I couldn't keep quiet. I once again appealed to

Kalinin, the director of P.O. Box A-1063. We met in early 1985, and I expressed myself on the subjects of Urakov and of proceedings at the institute. Urakov in turn declared that he couldn't work with me, and both Borovik and Tarumov entirely agreed with him. But their complaints were futile: Kalinin called on us to not "wash our dirty linen in public" and stated that a means of reconciliation had to be found. "You don't change horses in midstream."

Kalinin feebly attempted to get things back to normal, but the whole problem was very knotty, and the conflict between Urakov and me was too deep to be easily resolved. In general, Kalinin had to give preference to Urakov since he was head of the institute. Furthermore, around this time the Institute of Applied Microbiology finally received a permit to work on other, more dangerous agents, including plague, from the Third Main Directorate of the Ministry of Health, and Biopreparat lost any interest in tularemia. So the whole tularemia issue evaporated.

Still, I have never understood why Anisimov felt it necessary to invent this tale about the strain. It is probable that he had actually brought it in his pocket from Sverdlovsk but, fearing the reaction by his former directorate, either on his own initiative, or under advice from Urakov and Tarumov, he exchanged it for mine, thereby choosing the lesser of two evils.

I doubt whether Anisimov and Urakov were expecting this denouement, or they would have calculated their actions better. Incidentally, having arrived at Obolensk as a major, Anisimov was then promoted to colonel (his "unethical" behavior had no effect on his career).[3]

Since our military customer's parallel efforts to produce a genetically enhanced tularemia microbe at the Ministry of Defense laboratory at Sverdlovsk had ceased to be a secret, it was decided to ask the military to hand it over to us. But the talks dragged on. I was advised to continue work on the production of another new strain, which I had already begun before the Anisimov scandal. The military passed Anisimov's Sverdlovsk strain to P.O. Box A-1063 when our own work was coming to an end. Giving priority to an "alien" strain over ours, the Biopreparat management had returned it straight to the factory to produce a preparation for future experiments. They intended to use the military strain to produce the tularemia biological weapons, passing over our own Biopreparat production, which for some reason they assumed would be inferior.

This was my moment of triumph: it was found that the military's strain grew badly on regular nutrient media. In other words, it was not "workable," and in other aspects it was no better than our new strain (the T-14)! However, P.O. Box A-1063 warned me that Urakov and the military were trying to find some new fault with T-14, and that, in any case, we would not be given permission to use it. This is one more example (I have lost count of the number) of how they undermined our work. This example, and the countless others like it, only go to show that what really mattered in the branches of science controlled by the Soviet military was winning some perceived competition rather than achieving the stated objective.

I received great satisfaction from one other thing. The Customer, the military, had asked for and received a gene of the diphtheria toxin that we had cloned, the same one that was denounced as "useless" by Vorobev (see chapter 11). I had hoped to place this gene into the plague or tularemia agents in order to produce a bacterium with new and unusual properties, the stated goal of much of our research. Apparently the Customer shared my interest in this project, and attempted to test or to apply my results. Unfortunately, as usual, I was unable to find out anything from the Customer about the results of the work.

In spite of all this infighting, four of my colleagues received governmental awards for their execution of the five-year plan. In announcing these awards, Urakov added "a spoonful of tar to the pot of honey" by saying that "getting an award meant nothing, so that the award should only be regarded as a down payment." In fairness it is worth mentioning that Ryapis should have been among the prizewinners. His contribution toward the production of the strains must not be underestimated. But, unfortunately, shortly before this he had to leave Obolensk, owing to his daughter's serious illness and subsequent death. Ryapis had been a student of mine, and the director's hostility to me brushed off on him. Taking advantage of the opportunities afforded by the controlling department, Urakov then erased the name of Ryapis from all the work. This petty, spiteful action was typical of Urakov. Ryapis had done nothing to warrant Urakov's animosity. He had simply been my student.

As for me, in 1982, on the basis of my results over the previous five years, and for "services in the development of molecular biology and genetics," I was awarded the Order of People's Friendship. (The Decree of the Presidium of the Supreme Council was secret, so I don't know the exact wording.) Owing to my strained relations with the directorate, I was only given one of the less prestigious orders, but it is about the only one now that is ever quoted! Another paradox is that I was bestowed this particular award, bestowed for the "strengthening of peace" with which, in terms of my principal activity, I had had nothing whatever to do.

This is but one more irony of the Soviet system, where the means of protection from disease is a cover for offensive biological weapons research, and where a man who devoted years of his life to making deadly weapons strains suddenly finds himself awarded the Order of People's Friendship for the strengthening of peace!

NOTES

1. Matthew Meselson et al., "The Sverdlovsk Anthrax Outbreak of 1979," *Science* 266 (1994): 1202–1208.

2. Interestingly enough, this man was soon caught red-handed hawking something on the black market. I'm not sure whether it was vivarium materials he was selling, or feed for the animals. A vivarium is a laboratory that raises the live animals used in experiments.

3. Colonel Anisimov died in the mid-nineties. As the story goes, he was electrocuted when he climbed up a pole to illegally divert electricity to his dacha from municipal electrical wires.

CHAPTER 15

> To live means to struggle.
>
> Seneca

I have paid considerable attention to the character of Gen. Nikolai Urakov,[1] for otherwise it would be difficult to follow the whole course of events. In a closed city such as Obolensk everything depended on people like him. It is difficult to explain to those in the West what life in such a closed town was really like, and how dependent its inhabitants and workers were on such leaders. There were police stations on the two roads to the institute and adjacent to the town. Nobody could live at Obolensk without special permission. The town and its institute were carved right out of the wilderness; they could not even be found on a map. Every one of the institute directors was "god, emperor, and commander-in-chief." Not a single matter was resolved without them, whether domestic (flats, allocation of cars, weekend estates) or scientific, governmental, even Party-related business all fell under their purview. Unless they had displeased somebody higher up—such as a director of Glavmikrobioprom, or someone at the Military Industrial Commission or the Central Committee of the Communist Party (which seldom happened)—no one could get round these directors, including the local party bosses.

It was therefore interesting to see how Urakov solved his problems at party meetings. Since only a few party members were admitted to any secret business (maybe thirty out of about five hundred people attending), and since builders and laborers were predominant among the Communists, everything was said in figurative language, especially when discussing the results of any scientific work. Instead of "calling a spade a spade" as Urakov would in our own meetings, at party sessions he camouflaged our work by referring to our "newly developed vaccines," or new "methods of treating dangerous diseases."

He therefore had no opposition from the rank-and-file Party members, and thus he had a large numerical advantage when meetings were held. He used the same techniques for general meetings of the collective. At Party meetings, Urakov discussed different aspects of the "open" life of the institute: the mundane issues of construction, house building, labor, and financial discipline. Disputes or scandals among party members typically concerned domestic issues, most having to do with housing. Anything more sensitive was discussed in what we may call "Aesopian" language, using metaphor and fable. Urakov, just like anyone else, had no authority to disclose secret information even at party meetings.

Anyone who thought that the Anisimov affair of two years earlier in 1984 had been forgotten was sadly mistaken. Although some sort of truth had triumphed, and Anisimov had been given a reprimand for conduct "unbecoming a scientist," which was well known both at Obolensk and throughout Biopreparat, the tense situation at the laboratories did not subside. Though the work had begun to settle down, minor squabbles and intrigues still continued. But by now all attention centered on the new main building being constructed for the research and development of dangerous pathogens. The official handover of the project to the Institute of Applied Microbiology called for a whole mountain of safety instructions. In his attempts to speed things up, Urakov resorted to all manner of devices, including having false documents provided to the Public Health Authority. Officials of the P.O. Box A-1063 knew about these tactics, but they looked the other way, because completion of the first stage of building for the institute had fallen badly behind schedule, and the "higher-ups" were demanding that the building be certified as completed. The completion of this building was critical to the objectives of Bio-preparat and of our Customer, the military: safety considerations seemed secondary. The original design of the building specified that each floor, designated for work on one specific pathogen, was to be entirely self-contained, so that an outbreak on one floor could not spread to others. But, in his haste, Urakov cut corners. There was no point in objecting or arguing with him. Any haste in such an important matter was bound to cause trouble. And trouble was not long in coming.

Between 1985 and 1986 only five floors were close to being ready. We had to obtain permission from the Third Main Directorate to begin to work on the rest of the floors. But in fact even these five floors were not prepared for work on deadly pathogens. The inadequately prepared facilities were actually very dangerous. Nevertheless, Urakov insisted on having different documents prepared and signed for the directorate as if the safety precautions for the first five floors had been put into place. On the third floor, however, where work had already begun, there were two or three instances in which workers were infected with tularemia. Luckily no serious consequences arose. Fortunately, too, none of these infections were connected with my subdivisions under my control.

Two years had passed in our work since the Anisimov incident. And then thunderclouds began rolling up once more. It all began with the routine elections to the Academy of Medical Sciences. My regular competitors for these elections were V. Ogarkov and V. Vorobev, although both of them had only recently become corresponding members (the latter only two years previously).* This was too soon for a normal promotion: they could not have achieved anything substantial as scientists in that time to warrant such a career elevation. Furthermore, both of them served more as government officials than as research scientists at that time. Neither had done any work in the realm of open science. Instead of relying on achievement, they leaned heavily for their promotions on their vast networks of powerful contacts, which their lofty status in the establishment served to reinforce. I should mention that in those days no "special"

---

*See Appendix 1, "Academies."

places within the framework of the Problem "F" were allocated for elections to membership of the Academy of Medical Sciences (AMS) (or at least none that I knew of).[2]

On returning from Moscow one Tuesday I learned that the previous day a meeting of the Academic Council of the Institute of Applied Microbiology had been called, at which, at the insistence of the director, Ogarkov and Vorobev had been unanimously nominated for full membership in the AMS.

I also needed to be nominated for election to any of our academies. Both institutional and individual nominations are necessary. (The individual nominations must come from other academics.) The matter was therefore put on the agenda for the next Academic Council. The council could not refuse me, but Urakov took advantage of this delay to settle accounts with me and to rescue his comrades-in-arms from their competitor. He had motives beyond his general envy and dislike of me. Both Ogarkov and Vorobev were his superiors as well as his friends, and he still had not forgiven me for the Anisimov debacle. Having read out my application to his colleagues, he cast doubt on my scientific merits, brought up the "strain affair," and put the matter to the vote without any discussion. Contrary to common sense and in the light of Urakov's blatant massaging of the facts, the decision of his "colleagues" was unanimous. I did not deserve to be nominated. A suitable characterization was cooked up and signed by the director, the secretary of the party bureau, and the chairman of the local committee (without which no matters could be considered in the academy). Besides the statement that "in four-and-a-half-year's work at the Institute of Applied Microbiology Comrade Domaradskij has failed to display the necessary organizing abilities for a manager of this rank," it indicated that "in his activities Comrade Domaradskij has tended to exaggerate and distort the results of his scientific work." The absurdity of Urakov's action was so plain that Kalinin, head of P.O. Box A-1063, who had no special sympathy for my rivals, issued a new report about me, which followed the first one into the AMS.

The elections were held shortly after. None of the three of us was made academician, and the vacancy was closed! Clearly Ogarkov and

Vorobev's vast network of contacts had not been enough to overcome their lack of scientific merit, and the negative reports about me made my election impossible. The position was closed, apparently, to put an end to the whole troublesome matter.

My patience was wearing thin so I decided to go all out. This was a risky gamble, since at the time any complaint about one's superiors was likely to be passed over by the authorities to whom it was addressed, and moved up to the very superiors being complained of, which could hardly improve the situation of the complainant. But I had had enough, and was willing to take the risk.

In the Politburo we were then under the control of L. N. Zajkov, to whom I addressed a memorandum. In it I tried to paint an objective picture of the institute, calling attention to the unsatisfactory progress in the execution of our tasks, which I felt was largely due to the negative influence of the director, Urakov. I insisted that both the institute (because of its director) and to some extent Biopreparat as well were heading down a wrong and dangerous course. In my opinion, the principal error was the lack of any sound approach toward solving the fundamental issues of Problem Ferment. Without a thorough grounding in the basic sciences of molecular biology and genetics, we had no ability to produce the genetically altered strains demanded by the military.

I knew what I was in for. To submit such a paper to an official at so high a level necessarily meant that a great many people were going to see it, including my immediate superiors. So a complaint indirectly involved top management. At the same time, I was well aware of the course of movement of such papers: they were usually returned to the relevant departments with an order "to carefully investigate the heart of the matter, take any appropriate measures, and report."

But the hope was burning inside me that things might perhaps take a different turn. This hope was founded on a naive belief in the strength of my numerous scientific, administrative, and social services, as well as on the fact that I had been present at the inception of Problem Ferment and had been appointed to a number of positions by the Politburo itself. I hoped that my complaint would generate a close and unbiased inspection of our work plans, which could not have produced the results Urakov promised and the military demanded.

In the beginning I almost believed in miracles. An assistant of Zajkov asked to see me. Following the usual practice, the meeting was timed for the late evening. Zajkov's assistant listened to me attentively, though from various hints I realized that he had had the time to make certain inquiries. The conversation therefore ended without resolution. "Why bother Lev Nikolaevich (i.e., Zajkov) with this matter? Unless you insist, it would be better for your minister to deal first with the case," said he.

This was standard operating procedure. Usually when any major political figure received a complaint, as a matter of course he delegated it to his assistant, who usually passed it on to a minister, so he himself would not have to bother with it. So I was extremely disappointed but not really surprised at this development. But as it happened, the matter did not end there.

In early 1984 the colorless Rotislav Rychkov had been moved from his post as head of Glavmikrobioprom, and quickly vanished from the scene. In his place the party appointed V. Bykov, who had begun as director of the BVK (Candida yeast-derived protein) Factory in Kirishi, not far from St. Petersburg[3] and had then become secretary of the town council of the party. While working in Kirishi, Bykov had somehow contrived to defend his candidate's thesis.[4] He was later brought into the Central Committee of the Communist Party. Becoming the head of Glavmikrobioprom in 1985, Bykov succeeded in getting it transformed into the Ministry for the Microbiological and Medical Industries, while Kalinin took up the post of a deputy minister. This move to ministry status increased Bykov's political weight: a ministry is a higher-level organization than a main directorate, and thus a minister has more power than the head of a main directorate. Bykov was able to do this because he had the support of party leaders through his membership on the central committee.

I was summoned to meet Bykov the day after my talk with Zajkov's assistant. Bykov began by reproaching me for having appeared at the central committee in the first place. I had to go over the whole thing again. But I saw immediately that Bykov had prepared for the interview. He laid most of the blame for problems at the Institute of Applied Microbiology on me. He accused me of spending too much time at the Moscow laboratory and of not helping the director of Obolensk! Since I could in no way agree to this, Bykov decided that the matter had to be settled by a commission and proposed Burgasov as its chairman. By that time Burgasov, never an ally of mine, had already lost his job as deputy minister at the Ministry of Health and had become a consultant to P.O. Box A-1063. He was therefore completely dependent on Bykov and Kalinin. I had learned too late of Burgasov's collaboration in this "independent" commission, and I knew that it was useless to protest to Bykov.

This commission, consisting mostly of members of P.O. Box A-1063, turned up unexpectedly at the institute after the working day had ended. But Burgasov, to my surprise, did not appear. Instead, the commission was chaired by one Vadim Barkov, who used to be a small fry in the Military Industrial Commission. Before that, he had worked under Burgasov at the Ministry of Health. As a counterweight to the supporters of Urakov (who must have been forewarned of the commission's arrival) there were only a few people present who were on my side that evening. Things had been set up so that I was certain that there was no point in relying on this commission to arrive at an objective conclusion regarding my complaints.

The first interrogations were addressed to people on the director's side and then, late in the evening, to those on my side. This meant that there were merely a few minutes for a conversation with each of them. I was called last.

It was difficult to discover anything from the reactions of the commission, although it all became clear in a day or two, when someone from the commission returned to Obolensk. Urakov convened a meeting of the academic council at which the substance of my letter to the Central Committee of the Communist Party was briefly explained. Next, preprinted questions were handed out to the members of the academic council. I remember ones like, "Are you aware of any instances of the exaggeration of results on the part of Domaradskij?" and "How would you assess the role of Domaradskij in the organization of scientific research at the Institute of Applied Microbiology?" In order to suggest greater "objectivity" a secret ballot was arranged, as a result of which the right answers were obtained from nearly all the questions (out of twenty-six people, seven of them were in my favor).

But I never discovered what the commission decided. Only much later two pieces of paper came my way. It is possible to judge from them what sort of decision had been reached. For the sake of interest I will quote the contents of one of them in full:

> To Com. Barkov V. I.:
>
> Having regard to the great experience of I. V. Domaradskij, to his important contributions to fundamental research ensuring the execution of tasks assigned by the Customer during the 12th Five-year Plan, and also the desire of Com. I. V. Domaradskij to move to Obolensk for the continuance of this work and to depart from managing the laboratory at the Institute of Protein Biosynthesis, I consider it reasonable for him to be employed in his existing duties, though in concordance with his knowledge on the basic subject-matter at the Institute of Applied Micro-biology.
>
> 19.12.86
> Klyucharev[5]

The second paper is an extract from a draft resolution signed by V. A. Sizov, the secretary of the commission:

> ... to call attention to the abnormal professional relations between N. N. Urakov and I. V. Domaradskij, which are having a negative effect on the situation at the Institute.
>
> - to note the absence of transparency (within the limits of the established order) in the scientific, the Party and communal life of the Institute;
> - to draw the attention of Com. N. N. Urakov to the absence of team spirit in the consideration of scientific questions (participation by the deputy for science and, where necessary, by the leading specialists);
> - to give Com. I. V. Domaradskij responsibility for directing all

genetic research being carried out at the Institute, and accordingly to arrange for a redistribution of subordination among the respective branches, and to relieve him of any other duties (such as the works on the plant defense from pests, i.e., vermin and other parasites).

- to draw the attention of Com. I. V. Domaradskij to his need to give more weight to scientific organization, to make stricter demands on his scientific colleagues, including in the design, recording, and analysis of experimental results.

According to these papers, at least, I had won. Though it seemed that the commission had been stacked against me from the start, and Urakov had done all he could to present me and my complaint in a negative light, the members of the commission supported me. My scientific record spoke for itself, and the problems inherent in Urakov's regime were apparent to anyone with any pretense toward objectivity. Further comment would be superfluous, as they say.

But that was the last I heard of this document, which clearly had not suited somebody. It was never confirmed. Therefore, despite this decision, things did not turn out as I had hoped. I was extremely annoyed by this squelching of the report, as it had been, against all odds, so favorable to me. Shortly afterward, I learned that Bykov himself had decided to take the matter into his own hands.

A few days later Urakov, all his deputies, the secretary of the party bureau, and I were summoned before Bykov. Burgasov, Kalinin, and Vorobev also attended the meeting. Bykov wished to hear them all and then make a decision. First of all he gave the floor to the Obolensk party. Again, as was so often the case at the Institute of Applied Microbiology, I was showered with a hail of complaints and accusations. But I came under particular attack by Tarumov, the man I expected to support me.

Though Tarumov and I had clashed in the past over the Anisimov affair, by this time relations between us had improved somewhat, mostly because his relations with Urakov had soured. One bone of contention was that, on inviting Tarumov into the institute, Urakov had promised him the moon, and, in particular, a flat in Protvino. But several years had gone by and Tarumov continued to live in an institute flat with his wife, who didn't have even residential entitlement.[6] In addition, he finally became irritated by Urakov's obstinacy and by all the minor squabbling. So Tarumov had decided to complain about Urakov. He went so far as promising to support me. Had we managed to stick together, things might not have gone in Urakov's favor. But Tarumov had the mind of a military man; he had been one since childhood. He began his career as a cadet, and learned the habit of submitting to his superiors. Now he was a subordinate of Urakov. It is perhaps not surprising, therefore, that he changed his mind at the last moment. As for Borovik, frankly it would have been ridiculous to expect anything else from him. He was, as I have said, an expeditious man, and one of his salient characteristics was his ability to be on good terms with his superiors, whoever they might be.

Then my turn came, and I expressed everything I had been storing up throughout my years of working at the institute. I stressed the sergeant-major-like tactics of Urakov, his efforts to convert the institute into a barracks, and his incompetence in many departments. I complained of Urakov's arrogant and aggressive behavior. Most of all, I emphasized Urakov's scientific incompetence, his disregard of fundamental science, his cutting of corners, his lack of rigor and thoughtfulness.

I concluded by insisting that as long as he was director the institute would never be worth much. On hearing this, Urakov positively jumped in his seat (obviously, nobody had ever spoken to him like that before, especially in the head office), while the rest kept quiet, awaiting the reaction from Bykov. Vorobev and Burgasov mumbled something indistinct, although the latter, as chairman of the commission, ought to have said something more definite. Kalinin in his usual way proceeded to appeal to us to find some means of accord; he reproached Urakov but offered me no support. There followed the decision by Bykov: "They will have to be separated. They cannot work together." He knew I was right in many respects. But he also apparently felt that he had to support Urakov as head of an institute. He may have hoped that, bit by bit, things would work themselves out.

And that was that.

I went on working at the institute. Because of my status they couldn't remove me, and it would be no easy matter to find me appropriate alternative work. But Urakov could not tolerate the situation. After all that had happened he couldn't just hide in the bushes and, having firmly resolved to get rid of me, he "made some new decisions." He could not fire me; he simply decided to make my life miserable.

Urakov's first effort was to raise the question of the salary I was earning. It was nearly above that of a director, and higher than his own, which he had long resented. But there was nothing he could do about my salary, which had been fixed by the Military Industrial Commission before I came to Obolensk. His second step was to force me to move to Obolensk, which was not yet completed, and from which it was difficult to get to Moscow. The reason he gave was sound in principle, namely, that everyone working on the agents of dangerous infections should live near the institute. It was hard to object to this, especially after he had set a personal example by moving to Obolensk himself.[7] But it was beyond my powers to live in the backwoods without my family and to keep up two homes. What was more important, I would have lost the Moscow laboratory. The distance between Obolensk and Protvino was sixteen kilometers, but I had my own car, and from Protvino it was easier for me to reach Moscow for the weekends to be with my family. The thought of moving to drab, isolated Obolensk was intolerable to me.

Furthermore, the entire situation in the institute was disintegrating, which hindered our work and affected not only me but my colleagues and students as well. During my time at the Institute of Applied Microbiology, I had prepared about twenty Candidates of Science, many more than all the other doctors. But life was a trial for my candidates; their nerves were frayed by the delay in the confirmations of their thesis topics; doubts were expressed about the scientific value of their work, and they were even afraid of complications in submitting their respective defenses.

Under the rules of the Secret Service of Organization P.O. Box A-1063 (Biopreparat) members of the institute could only defend their theses before that institute's Academic Council. After that defense, each dissertation would be sent to Organization P.O. Box A-3092,[8] which, besides everything else, acted as Expert Council for the High Certification Commission. After an obligatory examination of the theses, the Expert Council merely submitted its decisions on each one to the High Certification Committee without indicating the title of the work or disclosing its content. The same conditions also obtained in the other institutes, which were connected with Problem Ferment in one way or another. Even some members of the plague-control system, if their work was of interest to Biopreparat, defended their theses at the Institute of Applied Microbiology as well.

In such situations a lot depended on the director. As the chairman of the thesis defense council, he was (and the position still is) in charge of making a sole decision as to the acceptance of a thesis for defense, which could afford numerous grounds for refusal, and there is almost no one to appeal to in the event of a refusal. To show how this works, I will refer to the case of E. Ya. Amirov, from my laboratory at the Institute for Protein Biosynthesis.

Amirov had been the first to succeed in transferring genes to the plague microbe, using heterologous bacteriophages. This was a significant accomplishment because the plague microbe has no phages able to transfer genes. Amirov had written his doctoral thesis from this material, which, under the regulations, he was supposed to do at the institute. He had travel to there nearly every day. The completion of his thesis coincided with Urakov's arrival at the institute. Having read through the manuscript, Urakov declared that "the material is not extensive enough for a thesis, even at candidate level." No matter how hard I tried, I couldn't make him change his mind. Somebody had instilled it deep in Urakov's head that laboratory technology on any biological-weapons project must necessarily form the basis of a candidate's thesis, but a doctoral thesis should be based on bioweapons industrial technology, although everyone knows how hard it is in our day to confirm even laboratory technology, and biowarfare industrial technology usually involves no science whatever! This conviction of Urakov's was unshakable: he insisted that in order to earn a doctoral degree at his institute, the research must be directly related to the problems of weaponizing or delivering bacteria.

While I don't deny that some science or engineering must be involved in bioweaponization, it hardly seems to require the sort of scientific research or discovery normally associated with the granting of a doctoral degree.

After Urakov's rejection of his work, Amirov gave up trying and lost interest in it. His thesis was buried in the deep archives of the institute. I have always suspected that Urakov's brutal rejection of Amirov's really quite significant work was done just to spite me, since I was Amirov's advisor. I pity Amirov for his loss of priority, but Urakov had more than that on his conscience![9]

Because of the endless squabbles with Urakov and his petty objections to everything I or my students did, my final year of working at the institute was nightmarish. Construction of the first stage of the gigantic complex, which had fallen behind schedule, was pushed ahead. The transporting of existing laboratories and the organizing of new ones meant an enormous pile of paperwork. At that time Urakov had put me in charge of the department for Special Technical Safety, so I had to handle it as well. That department was responsible for preventing laboratory infections and dangerous accidents such as the anthrax explosion at Sverdlovsk.

All of this activity was carried out against a backdrop of increasing international tension over biological weapons. Rumors were circulating about the apparent arrival of an international commission to investigate possible violations of the Convention on the Prohibition of the Development, Production, and Stockpiling of Bacteriological (Biological) and Toxin Weapons, and Their Destruction.[10] This was very disturbing to us, though our institute in the 1980s was not directly involved in the actual testing of biological weapons. (About the military's activities during this period, of course, I cannot speak; I had not been directly involved with the actual work of the military labs for years.)

These rumors stirred the regulatory service to greater activity. Judging from conversations and newspaper articles, I believe that the commission actually arrived at Obolensk, but luckily when I was no longer there. Recalling now how various kinds of camouflage had to be knocked together before the arrival of any commissions, I don't envy those who had the job of preparing for these inspections. I think that, had

I not left, in view of my notability among "normal" scientific circles, Urakov would have shoveled all this work onto me.

Altogether I had had enough of Urakov and the situation at the institute. I began to press Kalinin to transfer me to Moscow. Kalinin offered me a choice of two jobs, neither of which could be called impressive. One was at the Institute for Protein Biosynthesis with my own laboratory that I ran "for public considerations," i.e., for free. The other was the All-Union Institute of Biological Instrumentation (AIBI), which he had founded in the 1970s. Of course he preferred that I choose the second alternative on the grounds that, with the arrival of Bykov, he had lost influence on the Institute for Protein Biosynthesis, whereas the AIBI was his own "patrimony," and he would be able to help me to settle in and keep me within the system. I still had good relations with Kalinin at the time, and hoped that I would not be completely detached, under his patronage, from my years of research on Problem Ferment. So I thought it over and agreed with him. After my usual vacation, in the summer of 1987 I left Obolensk. I have never returned.[11]

As I think back now about Obolensk, it seems to me that in some ways this was the worst—the most poorly run and the least productive—of Biopreparat's facilities. As I have mentioned already, besides the Institute of Applied Microbiology, the establishment of P.O. Box A-1063 controlled quite a number of major research institutes. I was familiar with the activities of some of them from documents that landed at the council department (e.g., plans, reports, etc.) or directly through my inspections. I also had some acquaintances among their members, with whom I conducted some joint research. Of course, not everything proceeded smoothly in those institutes, and the general situation was more or less similar to that of Obolensk. But there were some fundamental differences: institutes such as the Institute for Ultra-Pure Preparations in Leningrad, and Vector, the Institute of Applied Virology in Koltsovo, were directed by civilians—young people with academic training—rather than by the military. In terms of their methodology (I mean fundamental matters in molecular biology and genetics, etc.) both of these institutes were far superior to the Institute of Applied Microbiology, and had wide links to the "outside world," which they maintain to this day.

But when it came to solving "special" problems—i.e., the actual development of biological weapons strains—the Institute of Applied Microbiology was well ahead of the other institutes in the Biopreparat system, though the Obolensk expertise in technological developments had been largely imported by Urakov and company from the Ministry of Defense.

Today, with the present conditions of overall conversion from biological warfare technologies to peacetime pursuits, the Institute of Applied Microbiology doesn't fit in easily. Certainly the situation there is harder than at the other institutes, which haven't neglected research in fundamental science. I am not claiming to be a prophet, but, based on my analysis of the situation in the late 1980s, I warned Zajkov, a member of the Politburo of the Communist Party, twice—once on October 27, 1986, and then again on May 8, 1987. I also went to Bykov, the minister of medical and microbiological industry (the successor organization to Glavmikrobioprom, which, as we have seen, Bykov reorganized in the early 1980s), on January 16, 1987, not to mention Kalinin, representatives of the military, and members of the Military Industrial Commission (who incidentally had once been my subordinates at P.O. Box A-3092, the Department of the Council),[12] about the likelihood of such an outcome. I tried to prove to them that Kalinin's system, even then, needed radical changes and to demilitarize, particularly at Obolensk. The insistence on technology and the mechanics of weaponization precluded real scientific discoveries and success, such as we have seen from Vector under Academician Lev Sandakchiev, who, as a civilian, had a much greater respect for fundamental science, and consequently was able to achieve much more.

But in general, though the Soviet Union undoubtedly had developed biological weapons (the Sverdlovsk anthrax outbreak and Aralsk smallpox incident are proof), very little was done to develop a new generation of these weapons, as had been the original goal of the secret Interagency Science and Technology Council of Molecular Biology and Genetics founded in 1973. The main reason for this failure was the lack of talent of its leaders and their inability to learn new ways of thinking. They could not understand molecular biology and genetics and tried to put new ideas into the Procrustean bed of old conceptions. They were also too impatient: from the beginning of any project under their control to its completion there was only a very short period—far too short a time to make certain that all the difficulties attendant on creating a new, more stable, more virulent, poly-resistant, and vaccine-resistant organism suitable for weaponization had been successfully cleared up.

Venturing to judge the effectiveness of work done on Problem Ferment as a whole, I have to say that it has justified neither the hopes nor the colossal investment of material. Essentially nothing remarkable was ever produced, apart from individual, nonfundamental results. The blame for all this lies with untalented management, those heads of institutions who did their best to snatch as much out of it as they could, with their gaze firmly fixed on their own ambitions.

By the end of my period in Obolensk it was decided to discontinue the work on tularemia, so that my efforts went for nothing. In a sense, tularemia had been a stop-gap program from the beginning: the military, as I have said earlier, preferred to develop contagious rather than noncontagious infections, as these, after the initial assault, could be trusted to work on their own. Once the laboratory had achieved sufficiently high safety standards to permit work with plague, for instance, tularemia was no longer a high priority, despite all the years and the effort we had devoted to it.

The results of my work on tularemia were published in 1991. Evidently someone decided then that they no longer needed to be kept secret. It was still prohibited, however, to mention differences in the virulence of strains, and especially to point out that strains of American origin were the most virulent. Perhaps someone published this research to demonstrate that the Institute of Applied Microbiology was really engaged only in defensive research, that is to say, in Problem No. 5—in order to hide the true nature of its work. This was one more smoke screen thrown up to convince the West that the Soviet Union was truly not involved in biological weapons research.

As for the other departments, I am ignorant of their actual fate. Of course, some work has been published since, in the name of transparency, in order to explain to the world scientific community the special status of the institutes within P.O. Box A-1063, and their work on highly dangerous infections. Under the changed political situation in Russia it has become impossible to conceal this work any longer. However, the overwhelming part of the results obtained and the methods devised, which could still be of use and serve as a basis for future research, is now inaccessible even to its authors, and a great deal of material vanished during the preparations for international inspections. So whatever there was that is possibly good and useful in our research is now unavailable, both to the researchers themselves, and to the rest of the world.

NOTES

1. Despite the fall of the Soviet Union, Urakov was until May 2003 director of Obolensk: whether anything has changed much now at Obolensk, I can't say.

2. See Appendix 1 for a detailed discussion of the academies and their role in Soviet intellectual, scientific, and political life—as well as for an account of my personal experience with the politics of election.

3. This factory had been involved a few years earlier in a notorious outbreak of allergies among the local population.

4. The candidate's thesis, which does not exist in the West, is preliminary to a full doctoral thesis.

5. L. Klyucharev, the reader will remember, was once my deputy at the Department of the Council and later the head of the Scientific Department of Biopreparat. He was an honest person and held in high esteem at Biopreparat.

6. During the Soviet period every person had to have a registration or residence permit. Since Tarumov did not have his own flat, his wife did not have a residence permit and might be deported from Protvino.

7. Urakov had long intended to build a house at Obolensk. Eventually he did so; his house is palatial by Russian standards.

8. The Department of the Council, which I headed until 1982.

9. The fate of Amirov was horrible: he was killed during a fire at his flat.

10. Signed April 10, 1972, entered into force on March 26, 1975.

11. Except for a motor trip I took with my son-in-law, my coauthor, Wendy Orent, and her husband, in April of 2002. We drove through the complex, which is no longer guarded and seems about to fall down, though it also has some lavish homes, one belonging to General Nikolai Urakov.

12. In the early1980s my former deputy at the council, Oleg Ignatev, became the head of the Department in the Military Industrial Commission that controlled our work.

CHAPTER 16

RETURN TO MOSCOW

Oeum et operam perdidi (I labored in vain)

Plautus

Why would I turn down an offer to move back to the Institute of Protein Biosynthesis, where my department, which by then had been converted back to a laboratory, had existed for fifteen years? It is not easy to give a clear answer.

The golden age of the laboratory occurred in the late 1970s during the "reign" of V. Belyaev and my time as head of P.O. Box A-3092, the Department of the Inter-Agency Council.[1] Back then there were good opportunities for work, including an ample supply of foreign currency for purchasing the modern equipment and materials required for anyone engaged in molecular genetics. My lab had all that was needed for modern molecular genetics research: centrifuges, Geiger counters, equipment for electrophoresis,* imported nutrient media, and so on. Altogether, my laboratory was fairly well equipped in comparison with many other microbiology laboratories, and I was able to conduct research at a level like that of scientists in the West. In addition, my laboratory was

---

*An electronic technique designed to separate, identify, or measure, the components of a protein mixture.

essential to the work of P.O. Box A-1063: at the time, that system was only beginning to develop its own laboratory base. My lab was devoted to research and development of new genetic methods essential to produce the strains needed by the military. In addition, I supplied the military with modified strains of various nonpathogenic bacteria, including clones of the EV strain of plague. Moreover, the discovery of plague plasmids came from work with these EV strains. But the situation had suddenly changed with the promotion of Rychkov and my own transfer to the Institute of Applied Microbiology; I no longer received the funds needed to procure equipment and to maintain the lab at the same high level.

In addition, as the number of new laboratories in the system increased, this forced the abandonment of "special activities" at the Institute of Protein Biosynthesis. These activities were transferred to the Institute of Applied Microbiology, where the powers that be felt there was less chance of information leaking out. The Institute of Protein Biosynthesis was an open institute, as I have said before. Attempts to keep secret the direction of my research, in particular my work with the EV strain of plague, would have been difficult. The information could easily have leaked outside my lab. Control of information was much tighter at Obolensk.

The special status of my laboratory and its workers—which had us reporting directly to V. Belyaev rather than to the head of the institute—had been a constant irritation to the management of the Institute of Protein Biosynthesis. The higher-ups could not understand the purpose of my work, or why I gave no assistance to the BVK factories,[2] for which the institute was a scientific supervisor. The management repeatedly tried to spark my interest in the technological problems in their manufacturing of BVK, but to no avail. This further intensified their irritation with me. The pressure upon me increased, especially after Belyaev's death from cancer in 1979.

My colleagues began to feel this pressure as well. There began a steady reduction of staff for my department, which resulted in a loss of three-quarters of the personnel. Eventually my department was reduced to a laboratory that looked very much as it did before I had taken it over.

Bykov's arrival further weakened the laboratory's position. Like Urakov, Bykov insisted that it was "distracting" me from my more important work at Obolensk. (He probably had no particular convictions on this score himself, but wished to support Urakov in his vendetta against me.) Bykov ignored the fact that most politicians, himself included, held several posts at the same time. Unlike me, however, they were paid for it, and received other benefits as well.

While working at the Institute of Applied Microbiology for nearly six years, I was unable to visit my laboratory more than once or twice a week. This couldn't but affect its work. Several people, fearing the management's displeasure, quietly began to deal with subject matter of the Institute of Protein Biosynthesis, which was of no interest to me. I had tried to keep the laboratory focused on general problems of genetics, instead of the applied bioweapons research of Obolensk, or the protein manufacturing issues of the Institute for Protein Biosynthesis. This abandonment of the original purpose of my laboratory on the part of some of my lab collaborators provoked a lot of clashes and quarrels among the staff. I had six or seven important staff members, each seeing himself as a prospective head of the department. The loss of Biopreparat's support for my laboratory further weakened my position. Finally, Kalinin, who had supported my laboratory, lost his influence over the civilian sectors of Glavmikrobioprom. The heads of the Institute of Protein Biosynthesis, eyeing my well-equipped laboratory and its superb staff, and seeing it no longer guarded, as it were, by Kalinin or Belyaev, decided to take it for themselves.

After returning to Moscow from Obolensk, when I managed to come to the laboratory at the Institute of Protein Biosynthesis, I had every hope that I would be able to keep it. But when I saw that it had completely collapsed, all my hesitations left me. Without strong backup I felt unable to start a new battle. I turned around and walked away. No one tried to stop me or suggested that I change my mind. I have never seen anyone from the laboratory since.

My new appointment at the All-Union Institute for Biological Instrument (AIBI), where I was appointed head of the laboratory "21-3," was a typical technical institute for the development of equipment for the rapid detection of microorganisms. Though it was under the control of Biopreparat and its secret service, the true purpose of its work was kept hidden. As a result, labs had only numbers. My lab was in Department 21, and its number was 3—thus 21-3. There were very few people with a biological or medical background and hardly any doctors of science. My own chief was only a Candidate of Science! Engineers, designers, and workers formed the bulk of the staff.

I therefore began to think over the future direction of my work so that I would fit in at the institute. Eventually, I chose, as the most versatile method for the development of rapid diagnostic tools, molecular hybridization. This technique could be useful, I felt, for the rapid detection of pathogens; as such, it had both military and defensive applications. It could be used in the event of a disease outbreak of unknown origin, to rapidly detect the nature of the pathogen and to enable rapid treatment of the infected.

First I needed to create a variant of the method suitable for a hardware configuration, which did not take long. I decided to base the method on the haptenization of nucleic acids (DNA) and to use suitable antibodies for detecting hybrid molecules.[3] This seemed to be an especially promising approach, since the AIBI had already developed an instrument for evaluating the results of immune-enzyme reactions.

Unfortunately, the conditions for performing the immunochemical part of this work didn't exist at the institute. The most serious problem was the lack of laboratory animals needed for this work. The faculty had no place to keep them. Furthermore, there were no facilities for the microbiological aspects of my research.

But a fortunate chance came to the rescue. Ryapis, who with my assistance had become head of the laboratory in the Faculty of Epidemiology at the I. M. Sechenov 1st Medical College, kindly offered me part of his premises and apparatus. As a result, and with the consent of V. D. Belyakov, the head of the Subfaculty of Epidemiology and an academician of the Academy of Medical Sciences, I obtained some quite tolerable conditions. There was no secret regime at the Faculty of Epidemiology, and my colleagues and I were able to work there quite freely. For my three colleagues and myself the epidemiology faculty was also a sort of window to the outside world, very different from the closeted conditions at the AIBI, where we spent most of our time. Our work at the AIBI was restricted. An elaborate protocol of secrecy was enforced. No one could leave the building without permission from the management, and it was forbidden to invite anyone in from outside. And all this at an institute where I was not even working on producing novel strains for eventual use as biological weapons, but only on methods of rapid detection! So, being able to work part-time at the Faculty of Epidemiology let us feel that we could breathe freely, as it were, for a time.

One of the worst things we had to suffer at the AIBI was the lack of a library. The most essential literature, mainly in Russian (reference books, dictionaries, teaching tools, and the like) was kept in an underground room far from the main building, and was effectively out of reach for most people. As for foreign literature, it was regarded as "nonrelevant" and was stored in a special room to which access was strictly limited. Foreign journals were considered to be nonrelevant if their titles were unconnected with any technology. This therefore ruled out all the journals on biology and medicine, even including *Nature* and *Microbiology Abstracts*. To those who are not familiar with security procedures, I would explain that these restrictions, according to their creators, were supposed to prevent outsiders from determining the true nature of the work of the institute. (By "an outsider" they meant anyone who was not directly engaged in the respective subject matter!) The secret service considered that the mere presence of literature on genetics and bacteriology could disclose the real purpose of our work. This pathological fear of discovery, which so hindered our research, was another of the absurdities of the Biopreparat system. The secret service at AIBI simply copied the regimes in place at Obolensk or Vektor, the virological laboratory at Koltsovo, though at AIBI the regime made even less sense, since the restrictions at Obolensk and Vektor were set up to preclude espionage. The result of all this secrecy and security at AIBI was that only a small number of people read anything at all. It was even worse than Obolensk. One could claim, in defense of these draconian security measures, that the members of the institute made use of the municipal libraries, though I never met a familiar face in any of them over many years.

Furthermore, despite the existence of a plan for "open publications," in the usual periodical journals or in books, our researchers published very little. This was true not only of workers at the institute, but also of those at other institutes within P.O. Box A-1063. Open publications were supposed to "cast a shadow on the wall," and show the world that we were not engaged in "secret" research. Sometimes the results of biological weapons research were also published, but in the form of "legends," in order to present some sort of explanation for the direction of research at our institutes, and also to disorient our potential enemies. The publications of the Obolensk institute on plant protection are one example of these legends, as were the placards in Serpukhov praising us for our work in developing novel vaccines. The problem was, of course, that the truly secret research could not be published, and for most people this necessitated developing publishable material unconnected to the main subject of work. Hardly anyone did so. On the other hand, many scientists are not fond of writing, and Kalinin's system never hinged qualifications or promotions on how much a scientist published.

On the other hand, all reports included all those who had contributed to them, and were therefore considered to be "publications." There was thus an overabundance of information contained in them for the required yearly certification—an internal review intended to make certain that each scientist was a genuine contributor to the institute. When leaving the institute, whether for retirement or for another position, members would be given documents stating the number of their "works," so that nobody could tell what any particular individual had been engaged in. The same applies to patent applications, the title page of which indicates the surname of merely one of the authors (and without the name of the invention given).

For my own part, since for years I had the lab at the Institute for Protein Biosynthesis, I had the ability to carry out genuinely open and publishable research. I was even able to publish several papers abroad. However, every time before I submitted an article for publication, I had to show my manuscript to the Biopreparat secret service. Furthermore, I remind Western readers that under the Soviet regime all publications, not just scientific ones, were passed through government censors. But that general censorship was not enough for the Biopreparat secret service!

After I lost my laboratory at the Institute for Protein Biosynthesis, I sometimes had to publish my own "nonrelevant" articles under the names of the other institutes, which came as a surprise to many of my friends there, who knew well that I had no direct connection with any of these institutes. However, it would have been inadvisable to explain these "anomalies" to them. I was forbidden from explaining my situation even to my friends at the institutes.

I have known a great many young people who were afraid to work in these closed scientific systems and who wouldn't even take the various perks and inducements offered to secure their acceptance of a position. They feared the obstacles to publishing, the loss of their scientific reputations, and the restricted travel abroad. Some of them refused big money, extra leave, and private flats as a matter of principle.

I had grown used to life in closed systems, even though I chafed sometimes at the restrictions of the AIBI. Still, life at the AIBI was tolerable, and I took every occasion to go off to the subfaculty of epidemiology, or to ask permission to visit a library (which was usually granted).

With the help of some good people I managed to assemble the necessary instruments and materials for my molecular-biology research. Another windfall was the fact that the AIBI had huge stocks of unused imported reagents, some of them very scarce and expensive. There were likewise no problems securing the needed research animals, since I had the resources of the Faculty of Epidemiology to aid me. Unfortunately, the quality and living conditions of these animals left a lot to be desired.[4]

Within a short period of time all of these factors helped to solve the research problems I was working on. We developed a method of molecular hybridization, which could be considered accessible over a wide range of practice. Now we had the method, but then we lacked the molecular probes, without which the diagnosis of infectious diseases could not be performed. The production of probes is a very complicated matter, involving considerable knowledge of the genetics of the appropriate agents, and a thorough understanding of the conditions necessary to work with them. Molecular probes are extremely valuable for the detection of natural or genetically altered disease agents. The probes are comprised of fragments of DNA of the microorganisms to be identified. The method of molecular hybridization is very sensitive and specific and now widely employed. But in the early 1990s the technology was still quite new, and it was necessary to use radioactive isotopes to estimate the results of the probes. This meant that the probes could be applied only in well-equipped and mainly academic institutes. To make things easier, I invented a method for the labeling of probes, which did not employ the use of isotopes and made the probes simpler and more convenient.

But when I began my work I needed highly qualified specialists in molecular biology, and I had none. It should have been possible to secure the staff I needed with a bit of cooperation from other institutes employing these specialists. But this was 1991. The Soviet Union was disintegrating; on December 26 of that year it would entirely dissolve. There had been plenty of these institutes in the old regime. But now, more than ever, competitiveness and interdepartmental fragmentation hindered our progress. Even the Bureau of Preventive Medicine at the Academy of Medical Sciences, the very body that approved the results of my research (by a resolution in January 1991), was of no help.

It was getting to the point where I wasn't able to secure the collection of bacterial "strains" from the Institute of Applied Microbiology, which I myself previously delivered from the Institute for Protein Biosynthesis (Urakov charged enormous sums for the extra strains that he once called "rubbish"). I couldn't even obtain the probe for the diagnosis of diphtheria, even though I had directed its design and coauthored the publication that discussed its development. I did, eventually, succeed in getting the probe from the institute, but I had to pay for it!

I had one other probe, which was suitable for the diagnosis of plague. But it didn't amount to much. We then proceeded to look for an alternative, and we hit upon oligonucleotide probes, which, according to the literature, could be biosynthesized. Luckily for me, the new, simpler method we had created, which did not call for nucleotides labeled with isotopes or expensive inserted enzymes, proved to be suitable to produce a plague probe as well. From the AIBI I secured the money for synthesizing two probes, but there was no money left for any more.

The question then arose about how to arrange the production of diagnostic kits for diptheria. In order to produce the kits, we needed to find customers willing to buy them. But, despite the worsening diphtheria situation throughout the country, and owing to the collapse of the economy, I was unable to produce the kits. The relevant documentation needed for any such technological process had to be approved by the Ministry of Health after extensive testing for safety and efficacy. All the paperwork was ready but it had yet to be used. Countless cases of diphtheria were left undetected because we could not fine anyone to pay for the diagnostic kits!

Alongside the hybridization, my laboratory was required to devise methods for the immunochemical determination of paprine and haprine (both varieties of the yeast derivative BVK), which were contaminating the atmosphere in the areas surrounding their production, and which was provoking, particularly in Kirishi, attacks by the "Greens" (a pro-ecology group of activists connected with the international Greenpeace organization). However, in this case we failed to adopt the method into practice; there were no funds available.

Meanwhile, the Soviet government and the socialist society in which I had lived were collapsing around me. After the Soviet Union had crumbled a period of total chaos ensued. This chaos also affected science and scientists, myself included. For two years (1992–1993) my entire income hovered between thirty and fifty dollars per month. Many institutes were closed, equipment was sold, and some institutes were privatized. Scientists had lost or left their work; they needed to search for any paying job, even one in no way connected with their profession. Some scientists, including many of my students, left Russia.

The deteriorating situation made it extremely difficult to continue to do research. Since everything had been sold, we had to buy the necessary materials, which once would have simply been supplied. In order to continue working, we had to engage in private microbiological and biochemical diagnostic work. Other institutes, such as the antiplague facilities and the Gamaleya Institute in Moscow, were even more financially strapped than we were.

I was still employed, but gradually I had less and less to do at the AIBI. I occupied some of my time in writing surveys, and a book, *Plague,* which was published in Saratov (1993), thanks to sponsorship by Prof. A. V. Naumov (who was director of the Russian Antiplague Institute "Mikrob"), and to assistance from Prof. G. M. Shub. I now admit that the book didn't turn out too well. There were no funds to obtain foreign literature and the Internet was not available to me then. I therefore could not keep up with the rapidly expanding foreign literature on plague. So the book had to be written using old data. Nevertheless I managed to express a few "seditious" ideas, and even to restore my primary role in various research areas that had been considered classified: the discovery of plasmids in the plague microbe; the production of a multiresistant EV vaccine; and the "chemical" vaccine of my student S. M. Dalvadyants, which consists of two antigens, one isolated from *Y. pseudotuberculosis,* and the other from *Y. pestis,* the plague agent. Dalvadyants based his vaccine on my work on cross immunity of plague and *pseudotuberculosis.* The vaccine is very effective for revaccination (after an initial use of the live vaccine strain EV). I also described my new hypotheses devoted to the existence of plague agent in nature. Despite its limitations, the book continues to be quoted in the scientific literature.

Besides the surveys and the book, I prepared lectures in biochemistry for the university students in Groznij, which I delivered there in April 1991. I derived no great satisfaction from this, since the level of student training in Chechnya, to put it politely, left much to be desired. I did my best to convey the subject, but at least a quarter of the students failed to comprehend it. Trying to encourage them somehow, I allowed them to choose the test questions themselves, but it did not help: the women, who made up the majority of the class, pored over the questions for two or three hours and then silently left the room. Of course, what could I expect, since most of the teaching staff were local trainees without advanced degrees in science? Before I came along, a pathologist had given the lectures on biochemistry and microbiology! In early January of the following year I was asked to go back there and hold a reexamination, but my wife wouldn't allow me. Disorder had already broken out in

Groznij. During the conflict the pro-rector of the university had been killed and his rector, V. Kankalik, had been abducted. Kankalik was never found. His efforts had served to support the college. He had even organized the new medical faculty.

Thus I spent the several years after I left Obolensk in work that gradually drew me further and further from the sphere of secret bioweapons research. The work I did after I left the system, which was related more to defense than to bioweaponry—and even my lectures in Chechnya—may be considered a peculiar form of payment for the years spent on biological weapons work.

NOTES

1. See chapters 8 and 15.

2. BVK is a protein of *Candida*, a species of yeast, grown on paraffins of oils. The protein is mixed with vitamins and then is used as supplemental cattle feed.

3. That approach is known as enzyme-linked immunosorbent assay or ELISA.

4. I had to use rabbits, and their maintenance was a problem. They had to live and forage outside. That made it very difficult to control the course of immunization and, therefore, the results of our experiments, especially in winter.

CHAPTER 17

FRIGHTENING TIMES

Before the phantom of their ideals
Many a warrior has bowed his head,
And shed his blood at the pedestals
Of a Golden Calf overthrown

Nadson Semen Yakovlevich (1862–1887)

W hile I was still at Obolensk, I began an effort to rehabilitate the reputations of my family members who had been tried and sentenced
under the Stalinist regime. This endeavor, which continued after I moved
to the All-Union Institute for Biological Instrumentation (AIBI) in
Moscow, caused me to think seriously about my life and to try to find an
explanation for many of my actions. I began to question, for the first time,
my adherence to a regime that had caused my family so much suffering,
and my collaboration in the biological weapons program. For the first time
in my life, I confronted the possibility of reviewing many of my actions,
of speaking freely, and of making decisions without hesitation. This self-
examination coincided with a disturbing time in Russia's recent history. It
was the beginning of perestroika, when one could engage without fear in
exonerating the innocents like my family members. Changes were taking
place in society, and I had to decide which shore to strike out for.

It was still not safe to speak out. Our secret service did not want to reconcile with the spirit of the times. Trying to clear my family members called for courage on my part, and a disturbing surprise for the regime. I did not think of the possible consequences; I decided to leave my active work on biological weapons research somehow or other.

The decision was almost effortless; it must have been in my blood. Disregarding the structure of the top-secret regime, my wife and I began attending almost all the demonstrations against the existing Communist government. I was searching for a way to express my attitude toward the Party. Later, when internal strife and jockeying for Party positions had reached their peak, I broke with the Party and sent them the following letter:

> To the primary CPSU organization
> at the A[ll-Union] Inst. for Biological
> Instrumentation from Member of the
> CPSU since 1946,
> Party card No. 01077679
> Domaradskij I.V.

Statement.

I became a candidate for membership of the Party during the last year of the War and I can rightly consider myself a veteran. Throughout the subsequent 45 years I have conscientiously performed everything required of me under the Constitution of the CPSU. Now, however, when there is no unity even in the Party and both its aims and responsibilities have become clouded, my conscience will not allow me to count myself a Member of the CPSU (especially as for the last two or three years my membership has taken the form of merely paying my dues).

In the light of this, I beg to withdraw from membership of the CPSU.

Signed: I. V. Domaradskij

On October 29–30, 1988, I attended an organizational meeting of the "Memorial" Society,[1] which impressed me at first by its aims and objectives. From beginning to end it was anti-Communist. Initially, the society was for victims of Stalinist repression and their relatives, but gradually, it became one of the centers of opposition. The society always opposed weapons of mass annihilation, including biological weapons, though too few people actually knew about them.

At Memorial Society meetings I met A. D. Sakharov, O. V. Volkov, and some other well-known upholders of social rights who had emerged from the underground. But I soon became disenchanted with the group. I have always felt that in the matter of rehabilitating the victims of Stalinist repression there ought to be no discrimination against the rights of anyone, and there should be no classification of people into "great" and "small." But the Memorial Society took an establishment approach: it ceased to talk of those who had suffered innocently, but rather spoke of people such as Bukharin and others, who, though they may have perished ultimately at the hands of the Stalinist regime, nevertheless played their part in initiating terror, collectivization, Lysenkoism, and the Doctors' Plot. They themselves had either directly or indirectly participated in preparing the ideological ground for political oppression, and they perished, not for principled opposition, but as a result of intraparty strife.

It is difficult, perhaps, for Westerners immediately to grasp the complexities of Soviet political life in the throes of destruction. The anti-Stalinist and totalitarian position of the former Soviet government did not have a single or clear-cut meaning. There are quite a few people who even now consider the anti-Stalinist course to be a serious error. At the same time they (I speak now of Gen. Pyotr Burgasov) believe that such persons as Lavrenty Beria, the dreaded KGB chief and notorious murderer, who was killed after Stalin died, to be victims of terror themselves![2]

My large family and I welcomed the political events of August 1991. We only feared that the reactionaries would win out and a new wave of repression would follow. The drama played out not far from us, and the tanks rolled right under our windows. We had to take the children out to our dacha to assure their safety. We monitored events at a safe distance. I was obliged to go in to work in the mornings, and in the evenings I witnessed the events at Russia's White House. Like many, my wife and I were worried about the confused situation and we were afraid that the state committee would use the state of emergency to gain the upper hand,

and that oppression would begin all over again. A great deal has been written about those days, but it is worth recounting what the atmosphere was like in an institute of the establishment such as the AIBI.

After the announcements on radio and TV and the subsequent "Dance of the Little Swans,"[3] I went off to work. The mood among many of my colleagues was quite gloomy, though some were genuinely glad and believed that the putsch would finish off the reform-minded Democrats. The roar of engines suddenly broke out under our windows at the All-Union Institute for Biological Instrumentation and one after the other, tanks and armored troop-carriers streaked down Volokolamskij Highway. When I returned home, armored troop-carriers were parked on either side of the bridge over the Skhodnya River. I could see others at the Borodinsky and Kutuzov Bridges near Moscow's city center and the Kremlin.

The next day, when some of my colleagues went off by themselves to the White House in response to President Yeltsin's plea for support, the management demanded an explanation from them in writing. During the putsch our chief forbade anyone to leave the building without permission; as former military people and opponents of democracy, they opposed the pro-government demonstrations near the White House. I quote here an extract from my colleagues' response, addressed to the deputy head of the department:

> We wish to inform you that our absence from the workplace on [21 August 1991] from [two o'clock on] was due to the critical situation which has arisen in our long-suffering country and in connection with [the] address by the President of Russia, B. N. Yeltsin, to his fellow countrymen.
>
> The victory for democracy has not been won by people sitting at their workplace or in heated flats, but by honest patriots who, in their country's hour of destiny, had answered the summons from the President of Russia.
>
> We feel insulted at such a time to be called to account for the fulfillment of our civic duties.

> 22 August 1991

G. I. Kondrashev
I. A. Apanovich
I. V. Semina
I. V. Ruzhentsova

Their unauthorized leave from the institute was considered a breach of working discipline and, as a head of the laboratory, I was held accountable for their actions. So, it fell to me to explain this action to the head of the institute:

> At your request I wish to state that from [three o'clock] on 21 August, at the summons of the President of Russia, I was in the building of the Supreme Soviet of the RSFSR.
>
> I entirely share the convictions of my colleagues. I am unable to condemn their actions, having no right to do so.
>
> 22 August 1991   I. Domaradskij

When the putsch was overthrown, the management promptly forgot about these letters, and no administrative action followed. But in private conversation management criticized me for having taken part in the events of August on the side of the "Democrats." The general mood of the institute was not on the side of Yeltsin's supporters, since the "hard core" at the facility was composed of military people, either on active duty or retired. I don't know how many active-duty military people were on staff (this was classified, though the director was certainly one of them). The rest did not conceal their army connections, and were proud of it (they were all graduates from Smirnov's establishment or from the chemical warfare division). So it is not surprising that Kalinin took care of his old comrades and allowed them to serve with honor in our institutions. Some of them were quite good specialists and they certainly pulled their weight, but most of them scraped for any job going, even those for which they were clearly unsuited, so long as it was not as duty men—janitors or porters—at the institute or at P.O. Box A-1063, or as storemen. Anyway, they were not expected to produce results. Their day usually began by reading the papers, mostly *Pravda* (The Truth) or *Krasnaya Zvezda* (The Red Star), swapping views on politics, arguing, or sometimes reminiscing about the days of Beria, when "complete order" prevailed. Was it any wonder, then, that on the first day of the putsch the Council of War Veterans and of Labor at the institute supported all the actions introduced by the State Committee on the State of Emergency?

I don't know how the August events were received by our other institutes, but I was told that Urakov had delightedly received the announcement by the committee, declared himself to be the commandant of
Obolensk, demanded absolute obedience, and stated that he would brook
no disorder. Knowing Urakov, I can well believe it.[4]

Still more unsettling to me than the days of August 1991 were the
events that unfolded at the Moscow White House on October 1993, by
which time I had left the All-Union Institute for Biological Instrumentation and was working at the Food Research Institute of the Russian
Academy of Medical Science (RAMS). Especially dreadful was the night
of the third and the morning of the fourth when the television screens went
black and all we heard were terse announcements on the situation surrounding the White House and the TV Center at Ostankino. The complete
silence from the authorities was particularly alarming; Yegor Gaidar's[5]
heroic speech in the square in front of the Moscow Soviet merely intensified our fear. The tension relaxed somewhat when the next morning the
tanks went past our house, and direct reporting began by American corre-
spondents on television. However, all day long, and even in the courtyard,
we could hear bursts of automatic fire and gunshots, and as before, there
was no means of knowing in whose favor the balance had swung.

By October 1993 the euphoria attached to Yeltsin and the move toward
democracy had started to fade. But the fear of the secret police coming to
power, and of the renewed threat of civil war, once again gripped a large
section of the public, though there was no visible sign of action on its
part. What was particularly striking at that time was the indifference (or
was it a wait-and-see attitude?) in the areas outside Moscow.[6] For my
part, the prospect of returning to the past brought nothing but foreboding.[7]

That dangerous time passed without bringing the secret police to
power, and without the outbreak of civil war. In many ways our situation
is now much better. But there are still many difficulties: in particular, a

low standard of living, corruption and criminality, (at the time) endless war in Chechnya, diminished political activity by the intelligentsia, and much more. What is happening now, almost ten years later, still does nothing to inspire confidence in a "bright" future.

NOTES

1. This group is more widely known in the West as the "Remembrance" Society, noted for maintaining the "Knigi Pamyati"—the Books of Remembrance. Relatives, friends, and others maintain those books in remembrance of persons who disappeared during the various purges, beginning with Lenin's 1919 purge and continuing until the death of Stalin (March 1953). (Ben Garrett)

2. Not long ago an unsuccessful attempt was made to rehabilitate Beria.

3. "Dance of the Little Swans" is a careless, light-hearted dance by Peter Tchaikovsky from the *Swan Lake* ballet. By playing this music on the radio, the insurgents wanted to reassure their listeners that everything was all right. TV and radio did not work during the putsch and after the announcements by the putsch's leaders the only thing shown on TV was this playful ballet. It was repeated several times and the expression "Dance of the Little Swans" became a synonym for the putsch. The Communist leaders always considered the people to be fools and this time they thought that such music might actually reassure the public.

4. Alibek wrote in his book *Biohazard* that Urakov apparently tried to commit suicide after the failure of the putsch.

5. Yegor Gaidar, economist and former prime minister under Boris Yeltsin, called on Muscovites to take to the streets to support the Yeltsin government against the attempted Communist putsch.

6. It appears to be true that the situation in this country is traditionally determined by events in the capital!

7. Speaking of the events of October, I cannot omit the mention of my student E. A. Yagovkin. Happening to be in Moscow (he lives in Rostov) on the night of August 4, he went on to the street and joined a small crowd of supporters of the current regime, which had gathered at the Yuri Dolgoruki monument. This famous statue of one of the founders of Moscow stands across from the Moscow Mayor's Office, near the city center and the Kremlin. Nobody could have predicted how things would turn out.

CHAPTER 18
AN INGLORIOUS FINISH

> Strength without intelligence
> collapses under its own weight.
>
> Horace

A fter the events of August 1991, the collapse of the country, which had begun during the period of "restructuring" and perestroika, proceeded at a rapid pace. This had an enormous impact on Biopreparat and the entire biological weapons research program. In 1992 President Boris Yeltsin issued a decree demanding that all development and testing of biological weapons agents cease, though in fact these weapons programs had already stopped when Vladimir Pasechnik, head of the Leningrad Institute for Ultra-Pure Preparations, defected to Britain in 1989. Meanwhile, the attacks on me at the All-Union Institute for Biological Instrumentation (AIBI), on the pretext of the irrelevance of my work and the lack of money for it, intensified. The management had not forgiven me for my open support of the democracy movement and of President Yeltsin; my "radical" ideas continued to annoy them. These attacks reached their climax in the early spring of 1992, when it was suggested to me—quite improperly, I might add—that I should retire. The "suggestion" was made by the head of the personnel department of the institute;

the official reason was my age. I was then sixty-seven years old, the retirement age of an ordinary worker, but not a senior scientist.

I protested and expressed my thoughts on the subject in writing to the director and to Kalinin, who had just then arrived at the institute. I told Kalinin that I regarded the "initiative" as an attempt to settle a score with me as a result of my open support for Yeltsin. Among other things, I said that I would not hesitate to use the media and my visibility to inform the public how the Russian Joint Stock Company "Biopreparat" had sprung up from the ashes of the former P.O. Box A-1063, and of what sort of work was done at Biopreparat and at the AIBI.

In his usual manner, Kalinin tried to calm me down and suggested that it was all a misunderstanding. He certainly did not support demands for my retirement. He implied that he did not want to lose me as a scientist. Perhaps he hoped that work on bioweapons would begin once more, and that I could be employed again somehow. He even promised to find the needed funds for my research. At the same time he stated that he would not be "put off by threats, as he had long since been shaken inside out." By this he seems to have been referring to some facts on which it is worth focusing attention.

I have previously mentioned Vladimir Pasechnik, the director of the Institute for Ultra-Pure Preparations in Leningrad, who was very close to Kalinin and well-informed about all the secrets of Problem Ferment. They were good friends, and Pasechnik, as director of the Leningrad Institute, reported directly to Kalinin as head of Biopreparat. But in 1989, while on a business trip to Paris, he refused to return home. According to an article by S. Leskov (1993),[2] ". . . such a sharp turn in the fortunes of Pasechnik is interesting in itself. As the son of a hero of the Soviet Union, he had become the director at the age of thirty-eight and he had excellent prospects of further promotion. But . . . , having realized the real objects of the programme (presumably Problem 'F') in his own words he had decided to reveal them to the world." And so he did: Pasechnik defected to Britain, was put in a safe house, and eventually told his British handlers all he knew of the Soviet biological weapons program, which was a great deal.[3]

The second fact concerns Kanatjan Alibekov,* Kalinin's deputy

---

*Ken Alibek.

director, who had taken Vorobev's old job as deputy director of P.O. Box A-1063 in 1987. A Kazakh by nationality, Alibekov gave the impression of being a very modest young man. I had first met him at Obolensk where he had come to defend his candidate's thesis. His supervisor was Vorobev. Nearly everyone had found this thesis "very weak," but allowance was made for the fact that Alibekov had written it at the remote station at Omutninsk, where he headed the experimental base. Conditions at Omutninsk were not promising for research: Alibekov was far from his scientific consultant, and there were bad conditions for his experiments. Later, "as a promising and energetic manager," he was moved to Stepnogorsk (in Kazakhstan) from which he was suddenly promoted to deputy head of P.O. Box A-1063. Alibekov's elevation to this extraordinary status had taken place remarkably fast, after just a few years. He arrived in Moscow with a completed doctoral thesis, which turned out to be essentially a simple industrial testing procedure. The basis of his thesis consisted of technological documents received from the Fifteenth Directorate of the Ministry of Defense. Alibekov used these documents for development of biological weapons at the new facility in Stepnogorsk. His was not a simple task but he accomplished it. His doctoral defense was held before a council chaired by Urakov, who passed him without reservation. But in the Expert Council, however, Alibekov's thesis met with some serious objections, mainly from the military. It was thanks to Kalinin's persuasion and Bykov's insistence that it was recommended for confirmation. Once again the youth of the candidate came into play, with some talk of how hard it was to prepare a thesis deep in the "sticks." Finally, they spoke of Alibekov's high rank, and of how he was responsible for all the science at P.O. Box A-1063. He was awarded his degree.

Alibekov subsequently directed the Research Institute for the Design of Applied Biochemistry. The institute designed manufacturing and testing equipment for biological weapons, as well as standards for industrial-scale manufacturing of these weapons. (Despite its public name, I think the institute was a legend, a cover.) I heard rumors that Alibekov even became a board member of a bank. Precisely how these and subsequent events developed I cannot say. I only know that Alibekov suddenly quit the military, left Moscow, and went off with his family to Alma-Ata in Kazakhstan. Later, he turned up abroad, where he revealed all he knew about Problem "F" to the world. And, as the deputy head of Biopreparat, he knew more than Pasechnik!

How Kalinin managed to keep his job after such an embarrassing defection one can only guess. Anyone else would have been sacked for misjudgments of a far less significant nature. It is likely that there was a lobby at work in the Military Industrial Commission (there is honor among thieves). In Stalin's time undoubtedly, and perhaps even later, Kalinin would have been at the least removed from his post and expelled from the Party. What or who helped Kalinin after Pasechnik's defection, I don't know. In any event, Kalinin was still a general, three or four years past the normal retirement age for Russian officers. How could Russia maintain that Biopreparat was solely devoted to peaceful research when its director, up until 2002, continued to hold a military rank? Moreover, another general, Valentin Evstigneev, a former head of the Fifteenth Directorate of the Ministry of Defense, worked alongside Kalinin for all those years.[4] He, too, only recently retired.

The third fact concerns V. S. Koshcheev, who was likewise involved in our Problem Ferment, though not as directly as the two previous men. A short time after he was appointed head of the Third Main Directorate under the USSR Ministry of Health he left on a trip for the United States. He never came back. I do not know all the circumstances of the Koshcheev defection, but he had a rank equal to a deputy minister and could be sent abroad on an official mission. Certainly as head of the Third Main Directorate he was in a position to know a great deal. We have seen throughout this book how the Third Main Directorate was intimately involved in all phases of research and development of biological weapons. Furthermore, the Third Main Directorate controlled a network of special hospitals and medical units to serve biological weapons research and development facilities. A second network of this directorate, known as "Program Flute," investigated biological agents that could cause both nonlethal and lethal organic and physiological changes in their victims. Several labs in this second network developed toxins and other substances for use against "individual human targets" as Kanatjan Alibekov, using his Westernized name Ken Alibek, puts it in his 1998 book *Biohazard*. Koshcheev took his state secrets with him. I don't know what became of him or where he is now.

As a result of all this, the business of P.O. Box A-1063 has become an open secret. Nevertheless I wish to draw attention to the incompetent organization of Biopreparat's work, which was supposed to be of great benefit to the country. It might have helped promote the status of biology, which had sunk so low during the reign of Lysenko and his minions. When the program arose, nobody could foresee its infamous finish, just as no one could have foretold the demise of the Soviet Union! But it seems to me that all might have been otherwise if Biopreparat had not scorned the fundamentals of science.

Having Stalin's support and after him that of other Soviet leaders, Lysenko hindered the development of biology and other sciences. For many years my country was isolated from the West, and as a result we were also isolated from scientific progress. Therefore, since the early 1960s many Russian scientists realized that, compared to the real progress made by Western molecular biologists and geneticists, we hadn't achieved very much. Officially, the government sought to close the gap between Soviet and Western science, particularly in molecular biology and genetics. To accomplish this a new system under Glavmikrobioprom was created, but the primary aim of the Biopreparat system was to use these disciplines to develop new types of biological weapons. The first task—closing the science gap—was a pretext for the second one, when it should have been the primary goal.[5] The elimination of this gap in the sciences was very important for the development of our national economy, but this aspect was forgotten by Biopreparat. I spoke about our goal openly many times on different government and party levels. The post-Soviet and present decay of Biopreparat's system is a consequence of this obliviousness.

I found a journalist, V. Umnov, who was interested in our research and was familiar with the subject of biological weapons in the Soviet Union. I wrote him a brief note offering my own account of affairs for the preparation of a suitable article for the *Komsomolskava Pravda*. How-ever, as usual, after my interview with Kalinin, I began to harbor doubts as to whether it was worthwhile sticking my neck out (a favorite expres-sion of my "colleagues" at P.O. Box A-1063), especially since I was still unclear about what I should say and to whom. It was, of course, one thing for Alibek, Pasechnik, and perhaps Koshcheev to publicize their accounts of the Soviet bioweapons program. They had all left the country. But I am a patriot and would never leave. Therefore, telling the truth entails many more risks for me, even now, under the new order. So I hesitated to speak to Umnov. The journalist tried to talk me through it, but he ultimately agreed to publish his article without any specific reference to me. Inciden-tally, in the course of his preparation he also used information derived from other sources.

Subsequent events took an unexpected turn. One day in late 1992 I was telephoned by a correspondent of *Newsweek* magazine, Carroll Bogert, who asked for a meeting. It came out that V. Umnov had told her about my note, and since she was involved in preparing an article for her own magazine on the development of bacteriological weapons in the Soviet Union, Ms. Bogert decided to speak to me. The subject of our con-versation was roughly the same as what she had learned from Umnov. In the article, which eventually appeared,* I was referred to as a "senior biologist" at one of the institutes of the Russian Joint Stock Company "Biopreparat."

The knowledge of our internal affairs by foreigners never ceases to sur-prise me. They set great value just on the maps they publish on the siting of the respective institutes and bases! I have always been skeptical that the documents could be kept hidden from other eyes. Putting myself in the position of any well-informed outside specialist, I have often tried to prove that the extremely clumsy attempts by our establishment, represented by Ogarkov, Vorobev, and others, "to cast shadows on the wall" could not fool anyone. But the prophet is without honor in his own country.

Neither these nor several other earlier articles produced any visible

---

*J. Barry, with Carroll Bogert et al., "Planning a Plague?" *Newsweek*, February 1, 1993, pp. 20–22.

reaction in the AIBI, although the *Newsweek* article had been translated into Russian. My involvement in all this was certainly discussed. One day I submitted a request to the deputy director of the AIBI, V. V. Buyanov, asking to declassify an old invention of mine that was necessary for publishing our method for hybridization. I was told that all the documents concerning our previous activities were eliminated. I was indignant. If this were true, they had destroyed a vast amount of data that could have been useful during conversion to commercial industry, not to mention the damage done to our science. Buyanov answered spitefully, shrugging his shoulders: "Who defended the White House and turned Umnov to our affairs?"[6]

The fact that documents relating to Problem Ferment have vanished has recently been confirmed under entirely credible circumstances. In the course of my work under the system at P.O. Box A-1063, I had joined the ranks of "restricted persons," meaning that I was forbidden to get an overseas passport without the Office of Visas and Registration (OVIR) stating any reason. I applied to the Administration of Archive Funds in the Federal Security Service, from whom I got the following answer: "Pursuant to the decision of the State Institute for Biological Instrumentation you are in possession of information which constitutes a State secret. Your application will be revised in 1998 only." I was left with only one avenue of appeal, the relevant commission at the Russian Federal Cabinet of Ministers. But for that purpose I needed a reference from the AIBI as to my "state of knowledge." It then emerged that all the documents that "incriminated" me had been "destroyed" five years earlier, and that, therefore, according to the director of the institute, there were no grounds for refusing me a passport! But I was not yet clear whether he would confirm this officially or whether a decision of Interdepartmental Commission would be sufficient. In any case, the irony of the situation was that their reason for refusing my passport was being covered up behind signatures asserting "nondisclosure" given many years earlier by representatives of a state that had long since ceased to exist.

I do not know what the scientific centers and institutes of the old P.O. Box A-1063 engaged in during the period of overall conversion from biological weapons research and production into vaccine production and other commercial ventures. Since 1987, I have had no contact with the facilities of the old P.O. Box A-1063. A few of my acquaintances left Obolensk after the collapse of the USSR. My information about the situation there is very scanty. I can judge the situation based only on what I saw at the AIBI. It is most likely that after President Yeltsin's decree dated April 11, 1992, the source of funding for biological weapons laboratories, which had seemed to be inexhaustible, dried up, and they had to look for ways to earn money. In the past, Biopreparat got its funds through the State Planning Committee (and the Military Industrial Commission). After the collapse of the USSR, it had to earn money by selling various medicines, reagents, and diagnostic kits, thanks to foreign investment for peacetime conversion, and to contracts with private Russian firms. Apart from that, Biopreparat receives only small sums from the Ministry of Health and others. The total funding available to Biopreparat facilities was (and is) considerably less than under the old regime.

But all this does not mean that the problem of biological weapons has been entirely forgotten. Several institutes, which would be quite appropriate for research and production of biological weapons, including those in the system of the Russian Ministry of Defense, have not been shut down, and it is hard to believe that they are engaged solely in defense matters. It must be remembered that in order to defend oneself one has to be in possession of what one must defend against. Otherwise, there is nothing to be done. Still, I can't say anything certain about the existence of biological weapons programs today. I do not know whether they still exist. No doubt, however, there are defensive programs, which are directed to combat bioterrorism.

Since the late 1980s it has become more and more difficult to "satisfy one's curiosity at government's expense," and earning money by whatever means possible is the rule of thumb everywhere. A variety of commercial bodies and foreign firms have sprung up, and they will pay far better rates than the institutes. Young people chase after either big money or "greens"—the Russian term for U.S. dollars. Most people are too desperate. They don't mind what they have to do, provided they get paid. Recently, one of my most promising colleagues scrapped his completed thesis and went off into insurance medicine, while another joined a guild of masons (not Freemasons!). Even my own family has not escaped. My eldest daughter was obliged to abandon an almost completed doctoral thesis and leave her institute for a foreign firm. The husband of my youngest daughter likewise ended his scientific career to seek a living wage elsewhere.

Inevitably, the value of scientific degrees and titles is much less than it once was. Even the holders of these degrees, not to mention the postgraduate students, are lapsing into a semidestitute existence. This is happening against a background of so many new temptations: you can buy anything you wish and go wherever you please. Those who still retain their attachment to science are streaming to the West. They include two of my own students (one has turned up in Canada, and the other in the United States). This national brain drain now appears to have subsided, although it has left very few capable young people.

The institutes are crumbling. Their premises are being rented for various offices or even warehouses, from which the considerable profits wind up in the hands of the top management. The gap between their earnings and the salaries of even "plain" professors is very large. Those institutional members who are actually flourishing serve as consultants, giving recommendations for the "adoption" of scientific developments into practice. There are no reagents, equipment is obsolete and unreliable, no animals are available for experiments, and foreign literature has ceased to arrive even in the central libraries.

The blame for all this lies with the government, although many of its members and of the president's entourage flaunt scientific decrees and titles. It is curious that no such attitude to science or to scientists was present under the old totalitarian regime. Even during the war and the period of postwar devastation, funds were available for the material support of scientists. There were special shops for issuing rations to them, and the Academy of Medical Sciences itself had risen on the waves of the war. While still a postgraduate student, I received support in the form of a monthly grant for the purchase of books.

Under these difficult financial conditions, there is the risk that some of our gifted scientists may take their knowledge and sell it to the highest bidder. On the other hand, I do not think that the actual theft and sale of biological weapons is much of a threat. I do not know how long the shelf life of weaponized agents might be, but since biological weapons consist of living microorganisms, I suspect that the period might have already ended. Furthermore, as I have pointed out many times, biological weapons are not strains, and to sell a loaded shell or a missile would be both extremely difficult and ultimately senseless.

In recent years a new and serious danger has arisen in the path of Russian science, i.e., the drain of our ideas to the West under the guise of agreements "to render material assistance to our scientists." Against promises of financial grants and scholarships, original ideas and fresh approaches to the solution of a variety of scientific questions are drifting abroad. Often they end up in the computer databases of those who distribute the funds. But they don't disburse much money. Instead, we face an unprecedented, and penalty-free, form of scientific espionage. It was well stated in ancient times: "I fear the Greeks even when they bring gifts."

The wide scope of research into molecular biology and the genetics of various microorganisms in foreign countries will enable work to be restarted (if it has ever really ceased there) on the development of biological weapons, but now at a higher level. Success in biological weapons development also depends on the results of fundamental research, the necessity of which I have stressed many times. This, ultimately, was the source of my trouble with the management of P.O. Box A-1063. Against a background of the decline of science in Russia, when money is only being allocated to enable the scientists to "keep their trousers up," and faced with a "brain drain"—as some of our best scientists drift away to the West and elsewhere—we shall never be able to "catch up." And if for some reason the West restarts any work on biological weapons, we are no longer in a position to meet that challenge with a program of our own.

NOTES

1. To this day I am mystified as to how Pasechnik found himself abroad; it must have been an oversight of our Secret Office (see chapter 8, "Renaissance").

2. S. Leskov, "Chuma I Bomba" (Plague and a Bomb), *Isvestiya*, no. 118 (June 26, 1993).

3. Vladimir Pasechnik (1937–2001) described the Biopreparat system at length to his British handlers. He was the first source to reveal that smallpox had been weaponized and that illicit stocks were maintained outside of the legitimate WHO depository in Moscow. His revelations prompted President George Bush and Prime Ministers Margaret Thatcher and John Major to challenge Mikhail Gorbachev about violations of the 1972 Biological Weapons Convention, which banned the production and use of biological weapons. Pasechnik's disclosures had an important influence on the later assessment of Iraq's ability to make bioterrorist weapons with knowledge obtained through its links with Moscow. After his defection, he worked for ten years at the Department of Health's center for applied microbiological research at Porton Down, Wiltshire, before forming his own company, Regma Biotechnologies, to develop new drug alternatives to conventional antibiotics; its first target was a treatment for tuberculosis. He died in 2001. (Ben Garrett)

4. He was appointed to that post after V. Lebedinskij's death in 1991.

5. See chapter 8.

6. I believe, in fact, that this document as well as others were only hidden and not destroyed.

CHAPTER 19
THE FINAL STAGE

This is the finger of God.

Exod. 8:19

As 1993 began, I found myself out of work for the first time in my life. I had never previously had to look for work. Given my background and the nature of the Soviet system, it had always found me. But in my present position it was not easy to make my own arrangements for work. Who wants a middle-aged man with a checkered and doubtful history, even one possessing two doctorates and having full membership in two academies? Many people saw me as a potential rival. As soon as I began to discuss employment, they found some reason why the job wouldn't suit me. No doubt some prospective employers were concerned about my radicalism.

In March 1993, having finally broken off with the establishment of the old P.O. Box A-1063, I moved to the Institute of Food Research of the Academy of Medical Sciences. The director of this institute, M. N. Volgarev, was canvasing for election as a full member of the AMS. His deputy, V. A. Tutelyan, wanted to be nominated as a corresponding member. Both were keen to secure any "vote." As I now suspect, their desire for my support was behind my new appointment.

Initially I was happy to take this position at the Institute for Food

Research. I liked the place; in principle I could engage in biochemistry and microbiology research there. Besides, as I have said, it was very hard to find a suitable position, and both the director and his deputy made many attractive promises. But as it happened, things were not as they first seemed. Although I had settled in, I had little to do. I lacked any sort of laboratory base or even any likelihood of future experimental work. I had just one task there: I had to solve a very important problem relating to the rapid detection of amanitin, a deadly mushroom poison. The problem was in no way connected with biological weapons. But I could make no headway in this effort because the institute had no money to buy the necessary reagents and equipment.

Altogether, to quote the ancients, "sic transit gloria mundi."* And if I still had any glory left, it was not much, though I always tried to do my best and observe the common decencies. Unfortunately, my sense of self-worth was not helped by the circumstances under which I had to live under the old Biopreparat system. I often had to think one thing and do another, which tended to make for a split personality accompanied by pangs of conscience. This was especially so during my life in Moscow and at the bioweapons research facility P.O. Box A-1063, for which, as I have said, I am forever seeking justification. While none of my new colleagues openly blamed me, some days I used to feel that they were pointing their fingers behind my back. Many people knew about my work at Biopreparat, but nobody understood its the exact nature. I was, therefore, the target, I believe, of much wild speculation, which was fanned by what administrators viewed as my unaccommodating nature.

Quite by chance one day I ran into Tarumov, whom I had known and quarreled with long ago in Obolensk. He had recently retired and settled down in a commercial firm. Willing to make fun of him, I asked, "Have you decided to atone for your sins in your old age?" He answered, "There's nothing to atone for, I was in the military. I did what I was told. It's you who need to atone. You went into it with your eyes open." I had to admit that he was right; nobody had forced me, although at the time of my move to Moscow and into "The System" they had not declared their hand. I suppose I thought that I could somehow change things and do

---

*"Thus passes away the glory of the world."

something useful for my country. I remain a patriot, despite everything that befell my relatives and everything that is happening now.

As for orders, it was not only the military whose members were told what to do. For Party members there was only one alternative: either give up the work and any chance of having a fulfilling professional life in science, or refuse to obey orders and be unemployable. Either of these would blight one's career, and not many people could face it (personally I can recall no such case).

My situation at the Institute of Food Research could only be called good in a qualified way. Since I did not have a laboratory, I turned to writing: besides the mushroom project, I had no specific amount of work I had to complete, so I could do as I liked. Nothing was ever required of me, nor could I seem to arouse any interest in anyone to pursue any particular scientific project. To find an interest meant you also had to find the money to pursue it, and the institute lacked money. After a lot of badgering, I finally procured a computer. I completed a first edition of my book on plague as well as a manual on microbiology in German (on which more below). I was aided in these endeavors when the library of the AIBI was shut down and a great many of the foreign journals, some of them very valuable, were passed on to me. This saved me a great deal of trouble, since I didn't have to run to the libraries every day.

Early in 1993 I received a proposal from the Germans to prepare a report on plague for the Bundesgesundheitsamt (their Federal Health Office), and I decided to try to write it in German. My attempt was a success, but owing to some confusion in Germany, my report was postponed indefinitely. Bolstered by my success with German, I delivered a course of lectures on microbiology at the Russian-German Free University in Saratov and prepared a book for students in German. The book was eventually published in 1995, though I was never to earn anything from it. The main point of writing it was my wish to offer substantial support in restoring some sort of cultural autonomy to those ethnic Germans in Russia who have not yet emigrated to Germany. In this I was also being driven by my nostalgia for the city of Saratov, which grows stronger as I grow older. I had lived there for a total of thirty years, and I witnessed the deportation of ethnic Germans at the outbreak of the war.

The normalization of science in Russia will be a long process. While the situation in Russia is gradually changing for the better, the same attitude toward fundamental research remains: funds can be found if one looks hard enough, but they are available for applied research only. And so it has always been the case, and there are never enough funds for equipment, reagents, and laboratory animals. Scientific work is still not viewed as prestigious, though salaries for scientists have become relatively higher, especially for members of the Academy of Science and the Academy of Medical Science. In the past, scientists had earned $40 to $80 per month, but as of January 2003, the salary was raised to $300. While the amount has improved dramatically, it is still a pittance compared to what Western scientists earn.

A large part of my life was taken up by a small private scientific research firm known as Ultrasan. I was employed on a part-time basis to study different aspects of the interrelation of microflora and the human organism, and the use of microbial metabolites, in particular, volatile fatty acids, for treatment of intestinal dysfunctions. I was able to write two large articles on this subject, both of which were published in a very prestigious journal. My work at Ultrasan remained sporadic as the need arose.[1]

Though occasional work at the Institute of Food Research and Ultrasan helped keep me busy, I still wanted to find a full-time job! In March of 1992, before I left the AIBI, I came across some notices in the papers about the demand in China for qualified personnel in my former line of work. I wrote to the embassy, offering my services. A year later, in May of 1993, having decided to move out of Moscow, I sent a similar offer to Kirsan Ilumzhinov, the first president of the Kalmyk Republic of the former Soviet Union. I received no answer in either case.

NOTE

1. Nevertheless, I am connected with Ultrasan to the present.

EPILOGUE

I said and relieved my soul.
Ezek. 33:9

I t is quite reasonable to ask, why have I written all this, since most, if not all, of the facts are already known? Despite the recent wide attention paid to the subject of bioweapons, I feel it is necessary to highlight key points that are usually ignored, overlooked, or avoided in the history of the Soviet/Russian biological weapons activity of Biopreparat. I also had hoped to unveil the reasons for our failure, despite our promising beginnings, to develop Soviet/Russian science in ways that could have enriched our molecular biology, our understanding of genetics, and our medicine. As I have repeatedly insisted, the major error the government made, and the reason for the ultimate failure of the Biopreparat system, is that biological weapons and their development were made the raison d'être of that system. Instead of trying to overcome the huge lag in scientific understanding brought about by the follies of Soviet science, Biopreparat focused on the development of a scientific program that brought no good to the country from among members of the international community or to anyone in the system, except the military and the scientists who worked for Biopreparat.

In the end, I had hoped to understand myself and my own motives, and to explain to others why so many of us have compromised our consciences and behaved contrary to our own views.

The first and second tasks are now completed, though I understand that my readers may well consider my observations as lacking some objectivity since I was not only a witness but also a participant in a vast number of events. Had I harbored any hopes of avoiding subjective judgments, it was hardly possible. Two aspects of my life are represented in this book that, unfortunately, did not cross. On the whole, I regret that I left my work at the antiplague system. I think I could have done more there, and I also regret that I could not keep in touch with the antiplague institutes after I left to work at Biopreparat. To me, the antiplague system was always the best, and the work I did there was my real work.

The cruel reality is that the system in which I spent twenty-three years of my life was a powerful current, sweeping aside any obstacles. I dared to stand in its way, but I overestimated my strength. Voluntarily or not, I have spoiled my relationships with many powerful people, and have finally found myself left with nothing. The people I struggled with are still in power, or at least in positions of significant influence. Almost all the officers of the former Military Industrial Commission, as well as Bykov and Kalinin remain in high posts in different government departments as they did before. And my nemeses, Vorobev and Burgasov, are firmly rooted now in the Academy of Medical Sciences.

Still, to say that I was left with nothing is perhaps an exaggeration. I have many students, including some who have achieved various levels of success in science. My relationships with them are of great value to me. I have been conferred many honors and titles, and I am a full member of two academies. Accolades aside, I have lost that which I have always aspired to: the possibility for active scientific work and a wide sphere of inquiry to stimulate my curiosity. I have a great deal of experience in scientific work, yet I no longer have the possibility of doing it.

My third task was to write this memoir in the hope of understanding of myself and my motives, and the compromises I have made to do the scientific work I love so much. That goal proved to be much more complicated than I ever imagined.

A great many people have now emerged who played a conspicuous part in the recent Soviet past, but are now distancing themselves from the former regime. I am referring to those who rose to the highest levels of power, or who actively supported those who did, and who considered themselves ideologists seeking to build a "new society." I had always suspected their sincerity. They couldn't help seeing where this idea of a society based on universal equality and brotherhood (i.e., Communism) was leading them, and still they behaved as if they truly believed in its possibility. What has been the motive behind their recent change of direction? Was it the crowd of disillusioned and impoverished people on the streets, and fears for their own future? In any case, having donned the robes of democracy, they persist in their illusion in order to stay in power.

Furthermore, those people who everywhere proclaim their sudden conversion from materialism to idealism, from Marxism-Leninism to religion, and who stand with candles in their hands in full view during the Eastern Orthodox or Christmas vigils are another great riddle to me. Their newfound religion merely demonstrates a complete lack of principle. Against this background one can't help respecting those apologists for the old regime who have remained true to themselves, or at least do not pretend to have decked themselves out in fresh paint. This is not to say that I am sympathetic to them or that I am supporting them. Many of these people are merely mourning for their lost privileges and their unfettered power over other people.

I don't think I lean toward any one party. I have had many occasions to wonder how my life would have turned out, had it not been for the events of October 1917. It is hard to imagine. However, dim shadows of the past have always hovered before me and have gradually caused me to take a critical look at my own actions and to seek some understanding. This concerned both my joining the Party, as I have described, and my activities with the Glavmikrobioprom system. I have always remained a patriot and I feel that everything I did was contributing to the public good. Therein lies the contradiction of my actions. Ever since my early childhood I have lived,

as it were, in two dimensions—one in real life, one in dreaming of another
. . . I therefore welcomed all the events of recent years. My sole regret is
that the burden of my own experiences often induced me to resort to half-
measures and held me back from taking a more active role in the democratic movement. I realize that I was only able to follow the path of least
resistance. I was not an active opponent of Soviet socialism and I did not
play any remarkable role in the fight against the Soviet power.

But there was one other reason that restrained me then and continues
to do so. I am referring to the vagueness and uncertainty of the ideas of
current leaders of the various parties and movements, and their inability
to reach a "consensus," as Gorbachev would say, in the struggle to build
a new society. The situation closely resembles the period between the
February and the October revolutions, which enabled the Bolsheviks to
seize power almost without a fight, with all that followed in its wake. You
would suppose that someone among the intelligentsia should have
reflected on this. For what took place I blame some of those who were
closest to me, with their attitude of yielding to force, although some of my
family did join the White movement. This attitude may have determined
the main course of my life. All I knew of the past was from hearsay, it was
too remote from me, and I could not stick to my principles, or what was
more, give up my career to join the ranks of dissidents. I have often
envied those people who found the strength to do so. However, while
paying tribute to their courage and being unwilling to insult them, I have
sometimes consoled myself with the thought that some of them found
themselves in very difficult situations, the likes of which have thus far
passed me by. This may be why a number of prominent fighters for
freedom are now among the ranks of either the extreme Right or Left.

At the same time I have never lost hope for the best or for radical
change. There were times when, in hushed whispers, I cast doubt on the
firmness of the existing regime, relying on the fact that not a single social
order has ever proved to be everlasting. This is not to say that I did not
sometimes genuinely believe in the rightness of my own actions. On the
whole, I think I was a fairly typical example of what Soviet power did to
thoughtful people.

In the end, I became a peculiar sort of dissident. Early in my life and career, I was too preoccupied with scientific problems and not enough with the motives of those who sought my help. Not unlike other Russians of my time, I always felt that I could not have changed the system; I could not win a fight against a totalitarian machine. Yet, I did what I could under the circumstances.

In spite of everything, I cannot feel repentant. I lived like most of my fellows; I was a product of my time. Besides, I tried always to avoid shabby acts and did a great many good deeds. I believe I have always been an honest man and, within the confines of the system that became my life, I always stood for what I believed in, often at great cost. I only regret that it is impossible to begin my life anew and start again in a Russia free from Communism!

I try to enjoy every day that passes and I put my trust in the support and understanding of those near to me.

APPENDIX 1

ACADEMIES

Much has been spoken and written on this subject, but it is worthwhile dwelling once again on certain details, especially since I have had to play a part in many of the events relating to this subject.

I recall that I became a corresponding member of the AMS USSR in 1969, when Z. V. Ermoleva had been largely responsible for my election. I now realize how difficult it is to get into any academy, especially from the provinces. Although it is never mentioned, other things being equal, the preference is given mostly to people from Moscow and St. Petersburg. Before my election to the academy and for some time thereafter I had genuinely supposed that the decisive role played in such elections was the scientific merits of the candidate. I subsequently became fairly good at playing the "academic games" and discovered that the main factors are not any scientific merit, but official standing and protectionism. This can easily be ascertained by analyzing what are called the assessments which are written on each candidate, or the scientific baggage of those who are elected. With few exceptions you will not find any "rank-and-file" scientists among the latter. Instead, you will see the surnames of the directors of academic institutes, their deputies, senior ministry officials, departmental people, and even publishers. In short, up to now the state academies (the RAS, AMS, etc.) remain as the preserve of the establishment. If someone has the good fortune to rise to the management of an academic institute he will automatically become a member of some academy or other; if as director then usually as an academician, and his deputy as a corresponding member. Therefore the halo surrounding each of them in the great majority of cases does not reflect their true worth. This had been the practice before the collapse of the Soviet Union; appointments to high positions and the allocation of places in an academy ultimately depended on the CC CPSU.[1]

Competition for places was stiff. Elections were usually preceded by a meeting of party groups composed of members of the academies and invariably attended by responsible workers from the CC CPSU. No more than two persons for each vacant post were customarily listed; the first-named was a creature of the Central Committee. Below the line were the names of other candidates. As many of the electors were Communists, it was not difficult to foretell the outcome of the ballot.

In both the "great" and "small" academies there existed another form of election, which Baev[2] used to call "election on trust." This form presupposed the allocation of special places for especially outstanding staff within the closed systems,[3] who had hardly any publications and were therefore unknown to the scientific community at large.[4] These vacancies were sometimes advertised in the papers, indicating a speciality which often bore no resemblance whatever to the actual job description for the applicant, which he usually did not get to know until the date of selection. During these elections, and of course after "consideration" by the Party group, the secretary-academician would suddenly announce an extra place, recite the merits of the applicant, and suggest to the "electors" that the candidate was a genuinely "worthy" individual. To outsiders the sudden appearance of new academicians or corresponding members in an academy caused no little wonder, although most of them took it as right and proper.

Shortly after the emergence of Problem Ferment this system was extended to us. Implementing it was simple, as the interdepartmental council was composed of responsible members of the presidia: Ovchinnikov (vice president), Skryabin (chief scientific secretary), and Baev—academician-secretary of a division of the USSR Academy of Sciences.

The Academy of Sciences therefore took on a number of members I knew at P.O. Box A-1063 and A-1968, in particular I. P. Ashmarin, creator of the Factor program, and V. H. Pautov, director of the military laboratory RIEH at Kirov. I remember especially the case of the "election" of Sandakhchiev, the youthful and capable, though still quite "green," director of the All-Union Research Institute at Koltsovo. He had not had time to defend his doctoral thesis before our council, as I was told on the phone by Belyaev, who nevertheless insisted that the thesis instantly be put toward confirmation. Belyaev also specified that, in three month's time, Sandakhchiev be awarded the rank of "professor." Following this a position was allocated, and he became a corresponding member of the USSR Academy of Sciences. This all took no longer than six months.[6]

Now I'm returning to myself. A post in microbiology for a full academician opened in 1980, but my competitor turned out to be Miss I. N. Blokhina, who was a deputy on the Supreme Soviet of the USSR, and a member of a Commission of Public Health at this council. What is more, she was the sister of N. N. Blokhin, the president of the AMS. Having not abandoned completely all faith in justice, I weighed my chances and entered the race for election against I. Blokhina, although before the actual election one of the academicians advised me to make a noble gesture and withdraw my candidature, which would "score me a good mark." I had miscalculated once again: I received only eight votes, and Blokhina got in. I had now made myself one more enemy. While she remained a deputy of the Supreme Council, and until the election of a new president of the AMS (it was V. I. Pokrovskij), she was a very influential figure. The matter was further complicated by competition from Ogarkov and Vorobev (I wasn't worried about the rest), who cooperated with Blokhina (Ogarkov had close ties both with her and her brother). My election was also opposed by Viktor Zhdanov, who could barely conceal his dislike of all those remaining on the Inter-Agency Science and Technology Council, from which he had been unwillingly removed in 1975.

All this played out in full view of the other members of our division, many of whom were on good terms with me. Lately I "failed" twice more; I had been up against Ogarkov and Vorobev, and, as the votes were split, the appointments lapsed. After the death of Ogarkov (1987) and the collapse of the USSR in 1991, Vorobev and I were finally "divorced." I was elected in "microbiology" and he somewhere else. A highly important part here was that played by Academician-Secretary N. Izmerov, to whom I can only be grateful. It was remarkable that once the obstacles had been cleared from my path I got twenty-two votes (out of twenty-six) "in favor" in the elections to the academy with one abstention, while Vorobev polled fewer! All of this may seem boring to outsiders, but it is very typical of the academies where even now there are some people with their hands on the levers of power who do as they please. In this context it is also worth recalling the election in 1991, in a speciality quite unconnected with the AMS (now the RAMS) of the wife of Lukyanov, the chairman of the last Supreme Soviet of the USSR, and author of the notorious "Declaration" during the events of 1991.[7]

I may seem biased to some people, but knowing how fate has dealt with many worthy scientists, I cannot believe that Madame Lukyanova could have got into this or any other academy without a "hidden hand." The same also goes for the former Minister Bykov at the Minmedbioprom, who well in advance proceeded to prepare himself a place in the AMS (in an extremely out-of-the-way specialization). However, in the last case the emphasis was on some major investments, supposedly for the development of "bioengineering," for example, in such concentrations of academic "votes" as the First Medical Institute after I. M. Sechenov. As it happens, Vorobev settled there after his leaving P.O. Box A-1063.

Against this background it was with a sense of great satisfaction ("the sixth sense of Soviet man") that most of the scientific community received the announcement of the creation of the new academies, which had become possible after the end of the USSR became clear and Russian sovereignty proclaimed. Out of the fifteen republics of the USSR, Russia was the only one not to have its own Academy of Science. The subject of the creation of a Russian Academy of Science has its own history, which is no longer of any interest. What is really important is the planning by a group of scientists headed by Academician A. M. Prokhorov, the Nobel Prize winner, for the organization of this academy—the famous seven "No's"—of which the basic principles were, in particular:

a. its independence of any state or other bodies;
b. dedication to the advancement of scientific thought, spiritual regeneration, and the raising of the standard of living and culture of the people of Russia;
c. democratization, demonopolization, denationalization of property, and integration with world science;
d. refraining from paying for the title of member of the academy out of the state budget.

The new organization also abolished the hierarchical principle of structuring the academy: preeminence was not to be accorded to the body of directors, but to the authors of discoveries and the creators of new directions in science.[8]

On August 31, 1990, at a combined session of the Russian Academy of Science, members of the three new specialized Russian Academies of Science—those of Engineering Science, Natural Science, and Agricultural Science—adopted the concept and elected a joint presidium headed by Academician Prokhorov. Without exaggeration, this session represented a landmark in Russian science.

Still, "man proposes but God disposes." Euphoria soon gave way to disillusionment when, after the collapse of the Soviet Union, the former USSR Academy of Science became "Russian"—just like the former AUS and the VASKhNIL,[9] which still kept the organizational structure and the old principles inherited from the old regime. Only two of the new specialized academies were retained, that of Engineering Science, which had recently seen raised to "state" level, and of Natural Science, from which the title of "Russian" had been removed (from being the RANS it became merely the ANS) and which exists as a public organization. This has certainly been due to the halfhearted and scrappy nature of all our recent political and economic transformations: the old bureaucracy, "repainted in three colors," still held power; the high scientific elite remained, unwilling to lose any privileges. Even the election of new members, well-known politicians, and public figures didn't help the ANS.

After a number of recent media attacks, the ANS stabilized. But the attitude of state academies toward the ANS remains unchanged: they don't notice it or seem to be surprised when they hear about it. Members of the ANS who want to run for candidacy in a state academy thus take off the symbol of ANS membership[10] and omit mention of their membership in ANS in relevant documents.

By now, however, the ANS stands firmly on its feet. It has more than one thousand members, who were once prohibited from being called "academicians" or "corresponding members," and who did not receive any salaries, but themselves provided material support to the ANS. It has more than twenty honored members, including Nobel Prize winners from various countries.

Certainly the working of the ANS suffers from a few weaknesses, the chief of which is the lack of money for the advancement of science. These weaknesses can be seen as "growing pains" and the result of the struggle for existence. Nevertheless, if life in Russia ever returns to normal and becomes completely civilized, the ANS has every chance of shining like a jewel.

NOTES

1. Central Committee of the Communist Party.

2. A. A. Baev was academician-secretary of one of the divisions in the USSR Academy of Sciences. See chapter 9, page 163.

3. The biological, chemical, and nuclear weapons systems.

4. See the discussion of "special places" in chapter 11, page 190.

5. Sometimes those co-opted into the work of these organizations, such as S. V. Prozorovskij, were also elected to an academy through this route.

6. I cannot be sure, but I imagine that Ovchinnikov had gone by the same route, having been only a candidate of science in the mid-1960s, and an academician by 1970! Skryabin likewise had a remarkable career. For many years, and

in spite of the existing tradition, he had performed the duties of senior scientific secretary of the USSR Academy of Sciences while being merely a corresponding member, which inspired the appropriate attitude towards him. Skryabin had several times "failed" his election to academician, and he became a full member only in 1979, but not without some pressure on the electors from Ovchinnikov and Baev. All three of them acted as one man, and you could be certain that if one of them said yes, the others would say likewise.

7. However, election to the academies was boosted not only by the men, but by their wives as well. One episode stands out. The wife of one candidate for membership of an academy was a fairly well known artist, and a great many influential people in scientific circles used to stay at their dacha. This lady had painted several of their portraits and shortly before the elections she arranged the varnishing at a very smart exhibition hall, which was a huge success. I have no wish to disparage the scientific merits of the artist's husband, though her contribution to his election as an academician could not remain unnoticed. This subject could be continued ad infinitum.

8. Putting this principle into practice, it was suddenly found that only a handful of full members of the USSR Academy of Sciences and the dependent academies actually satisfied these requirements, while new names got thrown up that hardly anybody had heard!

9. Lenin Academy of Agricultural Sciences.

10. This symbol, by the way, is very beautiful: it is made of silver, with a bas-relief of the famous geologist V. I. Vernadskij.

INSTITUTES
AND FACILITIES

INSTITUTES DIRECTLY INVOLVED
IN BIOLOGICAL WARFARE SYSTEM

Basic Organizations

1. Interagency Scientific and Technical Council under Glavmikrobioprom. (The original head of the Council was V. Zhdanov from 1973 till 1975; after him it was V. Belyaev, until his death in 1979. After Belyaev, the head was R. Rychkov; the last head was V. Bykov.)

2. Organization P.O. Box A-3092—Administrative (or special) Department of the Interagency Scientific and Technical Council under Glavmikrobioprom. (The head of the Department from 1973 till 1982 was I. Domaradskij, then Maj. Gen. D. Vinogradov-Volzhinskij.)

3. Organization P.O. Box A-1063—"Biopreparat." (Maj. Gen. V. Ogarkov was its head to 1979 and then for twenty years Maj. Gen. Yu. Kalinin.)

4. Organization P.O. Box A-1968—the Fifteenth Directorate of the Soviet Army. Its chief was Col. Gen. E. Smirnov to 1985, and then Lt. Gen. V. Lebedinskij.

5. Third Main Directorate of the Soviet Ministry of Health.

6. Special Department (parallel to Biopreparat) under the Ministry of Agriculture of the USSR.

Institutes and Facilities of Biopreparat:

All-Union Institute of Applied Microbiology (Organization P.O. Box G–8724), Obolensk, Moscow region.

All-Union Institute of Molecular Biology ("Vector"), Koltsovo (not far from Novosibirsk).

Institute for Ultra-Pure Preparations, Leningrad (now St. Petersburg).

Institute of Immunology, Lyubuchany, Moscow region.

All-Union Institute of Biological Instrumentation (AIBI), Moscow.

Scientific and Production Base, Omutninsk, Kirov region.

Progress Scientific and Production Base, Stepnogorsk, Kazakhstan.

Institutes and Testing Ground of Organization P.O. Box A-1968

Institute of Military Technical Problems, Sverdlovsk (now Ekaterinburg).

Research Institute of Epidemiology and Hygiene (RIEH), Kirov (now Vyatka).

Virology Institute, Zagorsk (now Sergiev Posad), Moscow region.

The Aral Sea, Vozrazhdenie Island.

Institutes of the Special Department under the Ministry of Agriculture of the USSR

Institute of Veterinary Virology and Microbiology, Pokrov, Vladimir region.

Institute of Foot-and-mouth Disease, near Vladimir.

Institute of Phytopathology, Galitsino, about 40 km west from Moscow.

INSTITUTES INDIRECTLY INVOLVED IN BIOLOGICAL WARFARE SYSTEM

Institutes of the Academy of Sciences

M. M. Shemyakin Institute Bioorganic Chemistry, Moscow.
Institute of Biochemistry and Physiology of Microorganisms, Pushchino-
    on-Oka, Moscow region.

Institutes of the Academy of Medical Sciences

N. F. Gamaleya Institute of Epidemiology and Microbiology, Moscow.
D. I. Ivanovsky Institute of Virology, Moscow.

Institutes of the Ministry of Health

All-Union Antiplague Institute "Mikrob" (now " All-Russian"), Saratov.
Antiplague Institute, Volgograd.
Antiplague Institute of the Siberia and the Far East, Irkutsk.
Antiplague Institute, Rostov-on-Don.

OTHER INSTITUTES MENTIONED IN THE BOOK

Institutes of Glavmikrobioprom Not Connected with Biopreparat

All-Union Institute Protein Biosynthesis.
Institute of Genetics of Industrial Microorganisms.

Institutes of the Ministry of Health

Stavropol Antiplague Institute.
Alma-Ata Antiplague Institute.
Institute of Clinical Immunology, Moscow.
G. N. Gabrichevsky Institute of Epidemiology and Microbiology, Moscow.

Antiplague System of Siberia and the Far East (in 1957–1964)

Irkutsk (institute).
Ussuryisk (station).
Vladivostok (section of Ussuryisk station).
Nakhodka (section of Ussuryisk station).
Khabarovsk (station).
Vanino (section of Khabarovsk station).
Yuzhno-Sakhalinsk (section of Khabarovsk station).
Chita (station).
Borsya (section of Chita station).
Kyakhta (section).
Kyzyl (station).
Gorno-Altaisk (section).

GLOSSARY OF NAMES

S. I. Alikhanyan. Geneticist, doctor of biological sciences, professor, director of the Institute of Genetics of Industrial Microorganisms in the sixties and the early seventies.

V. V. Akimovich. Doctor of medical sciences, professor of microbiology of the Saratov Medical Institute.

N. I. Aleksandrov. Major general, doctor of medical sciences, professor one of the institutes of the Ministry of Defense.

K. Alibekov. Colonel, doctor of biological sciences, deputy chief of Biopreparat 1987–1991. In 1992 emigrated to the United States. Author of *Biohazard* (1998).

E. Ya. Amirov. Candidate of medical sciences, research fellow of the laboratory of extrachromosomal heredity at the Institute for Protein Biosynthesis.

Yu. A. Andropov (1914–1984). Chief of the KGB. Then, after Brezhnev's death, the General Secretary of Central Committee of Communist Party of the USSR (1982–1984).

G. Anisimov. Colonel, candidate of medical sciences. Worked at the Sverdlovsk Militery Institute, and then at the Institute of Applied Microbiology in Obolensk (near Moscow).

A. M. Antonov. Pathologoanatomist, professor of the Saratov Medical Institute.

M. I. Antsiferov. Candidate of medical sciences, expert in field of tularemia. After my departure from Irkutsk for two years was a head of the Irkutsk Antiplague Institute.

I. A. Apanovich. Scientific fellow of my laboratory at the All-Union Institute of Biological Instrumentation.

I. P. Ashmarin. Major general, academician of the Academy of Medical Science, one of deputy chiefs of the Organization P.O. Box A-1968 (or the 15th Directorate of the Ministry of Defense).

L. A. Avanyan. Doctor of medical sciences, research fellow of the Stavropol Antiplague Institute (The North Caucasus).

A. A. Baev (1903–1994). Well-known biochemist, member of the Soviet Academy of Sciences, secretary-academician of molecular-biological section of the Academy, member of the Interagency Scientific and Technology Council on Molecular Biology and Genetics. Like many other Soviet scientists, till the sixties, for eighteen years he was in the prisons and the camps.

Evgeniya. E. Bakhrakh. Doctor of of medical sciences, senior research fellow of the "Mikrob" Institute (Saratov). She suffered during "Doctors' Plot" events.

Ya. L. Bakhrakh. Research fellow one of the institutes in Saratov. He suffered during "Doctors' Plot" events.

V. I. Barkov. Functionary of the Military Industrial Commission.

O. V. Baroyan. Well-known expert in epidemiology, academician of the Soviet Academy of Medical Sciences. In the sixties and the seventies was a head the N. F. Gamaleya Institute of Epidemiology and Microbiology of the Soviet/Russian Academy of Medical Sciences.

N. G. Belenkij. Biologist, took part in work of session of the Lenin Academy of Agricultural Sciences or VASKhNIL (July 31–August 7, 1948), adherent of T. Lysenko.

V. D. Belyaev (1918–1979). Chemist (one of creators of the toxic gas sarin), head of Glavmikrobiompom under the Soviet Council of Ministers, head of the Interagency Scientific and Technology Council on Molecular Biology and Genetics (1975–1979).

A. I. Berdnikov. Professor, the first director of the All-Union Antiplague Institute "Mikrob" (Saratov).

A. L. Berlin (1903–1939). Deputy director of the "Mikrob" Institute. In 1939 he was infected with plague in the laboratory. Unaware of his illness, he traveled to Moscow, where he died.

N. N. Blokhin (1912–1993). Oncologist, academician of the Soviet Academy of Sciences and the Academy of Medical Sciences, president of the Academy of Medical Sciences in 1960–68 and 1977–87.

Irina Blokhina (sister of N. Blokhin). Microbiologist, academician of the Soviet/Russian Academy of Medical Sciences. Head of the Microbiological

Institute in Nizhny Novgorod (former Gorky); the city is about 430 kilometers east of Moscow on the Volga.

Caroll Bogert. Correspondent for *Newsweek* magazine.

A. A. Bogomolets (1881–1946). Pathophysiologist, academician of the Soviet as well as the Ukrainian and Byelorussian Academies of Sciences, president of the Ukrainian Academy 1930–1946.

R. V. Borovik. Doctor of veterinary sciences, professor, one of deputy directors of the Institute of Applied Microbiology in Obolensk (not far from Moscow).

A. E. Braunshtejn (1902–1986). Well-known biochemist (discoverer of amino acid transaminases), academician of the Soviet Academy of Sciences and the Soviet Academy of Medical Sciences.

L. I. Brezhnev (1906–1982). General Secretary of Central Committee of the Communist Party of the USSR (C.P.S.U.) 1964–1982.

S. E. Bresler. Expert in field of molecular biology and genetics, opponent of T. Lysenko.

P. N. Burgasov. Ardent adherent of Stalin and his helper Lavrenty. Beria, major-general, epidemiologist, long worked at a system of the Ministry of Defense, deputy minister of Health (Chief Medical Officer of Health) 1965–1986.

A. I. Burnazyan. Lieutenant general, deputy minister of Health (head of the Second and Third Main Directorates).

Frank Burnet (1899–1985). Well-known Australian scientist, virologist, and immunologist (Nobel Prize in 1960).

V. V. Buyanov. Doctor of medical sciences, deputy director of the All-Union Institute for Biological Instrument.

V. A. Bykov. Academician of the Academy of Medical Sciences, minister of the Medical and Microbiological Industries from 1984 till collapse of the USSR in 1991 and head of the Interagency Scientific and Technology Council on Molecular Biology and Genetics.

G. V. Chuchkin. Chief of Biological Weapons Directorate of the Military Industrial Commission in the late seventies and in the early eighties.

S. M. Dalvadyants. Immunologist, doctor of medical sciences, senior research officer of the Mikrob Institute in Saratov.

V. Debabov. Geneticist, doctor of biological sciences, professor, after S. Alikhanyan (see above) was a director of the Institute of Genetics of Industrial Microorganisms.

Yu. A. Demidov. Doctor of low sciences, professor, colonel of police.

I. A. Deminskij (1864–1912). Expert in plague. He proved by his own death in 1912 that suslik plague was identical to human plague.

Shu-li Di. Parasitologist, Chinese expert in plague.

M. M. Domaradskij. First cousin once removed of my father, a victim of Stalin's terror.

V. V. Domoratskij (Domaradskij). My father.

G. I. Dorogov. Colonel of KGB, chief of secret service of Biopreparat.

O. V. Dorozhko. Candidate of medical sciences, expert in tularemia and plague, research fellow of the Institute of Applied Microbiology in Obolensk (near Moscow).

I. Ya. Drevitskij. My mother's grandfather.

Ya. I. Drevitskij. My mother's father.

S. I. Drevitskij. Uncle of my mother.

M. S. Drozhevkina. Expert in brucellosis, doctor of medical sciences, professor, deputy director of the Rostov Antiplague Institute (till 1965).

I. S. Dudchenko (?–1917). Founder of the bacteriological laboratory in Chita (the Trans-Baikal area). On the night of June 5–6, 1917, while preparing to travel to a plague focus, he and his seven assistants were savagely murdered by bandits.

B. Ya. Elbert. Expert in tularemia, professor. Together with N. Gaiskij (see below) developed a life vaccine against tularemia.

Z. V. Ermoleva (1898–1974). Academician of the Soviet Academy of Sciences. She was the first who after Englishman Alexander Fleming (the Nobel Prize winner of 1945) received penicillin and used it on the front during the war of the USSR with Germany (1941–45). She was also a great expert in the cholera problem. In the twentieth century she proved that water luminous vibrio is a cholera one. Her figure became a prototype of the heroine of the novel *The Opened Book* (Otkrytaya kniga) by the Russian writer V. A. Kaverin (brother of L. Zilber, see below).

V. N. Fedorov. Noted expert in field of plague, doctor of medical sciences, professor of the Mikrob Institute (Saratov).

B. K. Fenyuk. Noted expert in field of plague, doctor of biological sciences, professor of the Mikrob Institute (Saratov).

Yu. A. Filipchenko (1882–1930). Biologist, a founder of first chair of genetics at the Leningrad University (1919).

Ye. Gaidar. The first prime minister of Russia after the collapse of the USSR in 1991.

N. A. Gaiskij (1884–1947). Doctor of medical sciences, professor. From 1939 until his death in 1947 he was head of the plague control laboratory in the small town of Furmanovo in the Ural Region, served as deputy director of the Antiplague Institute in Irkutsk. Before this appointment he had successfully completed experiments begun before the war with B. Ya. Elbert to produce a live tularamia vaccine. Like many other Soviet scientists, he suffered persecution; in 1930 he was sentenced to five years in the camp.

K. G. Gapochko. Major general, professor, worked at one of the military institutes in Leningrad (now St. Petersburg).

N. Gaponov. Cousin of my mother.

Nina Gefen. Colonel, microbiologist, doctor of medical sciences, professor (wife of N. Aleksandrov, see above).

N. N. Ginzburg. Noted expert in immunology, doctor of medical sciences, professor. In the forties, working at the Military Institute in Kirov (now Vyatka, northeast from Moscow), he developed a life vaccine against anthrax.

O. S. Glozman. Pathophysiologist, doctor of medical sciences, professor of the Saratov Medical Institute.

Glukhov. Chief of personnel department of the Institute of Applied Microbiology in Obolensk (near Moscow).

D. M. Goldfarb. Noted geneticist, doctor of medical sciences, professor, disabled veteran of the war with Germany (1941–1945), dissident.

E. P. Golubinskij. Biochemist, doctor of medical sciences, professor, head of the Irkutsk Antiplague Institute since 1979.

A. A. Grechko (1903–1976). Marshal of the Soviet Union, Minister of Defense 1967–1976.

A. G. Greten. Expert in histology, doctor of medical sciences, professor of the Medical Institute in Nizhny Novgorod (the city is about 430 kilometers east of Moscow on the Volga).

I. C. Gunsalos. Well-known American biochemist and microbiologist.

A. Heinaru. Expert in molecular biology; reader of the University in Tartu (Estonia).

K. Ilyumzhinov. President of the Kalmyk republic (in the Northeast Caucasus).

N. N. Ivanovskij. Biochemist, professor of the Saratov Medical Institute and head of biochemical laboratory at the Mikrob Institute (in the early fifties).

N. F. Izmerov. Expert in occupational hygiene, academician-secretary one of sections of the Soviet/Russian Academy of Medical Sciences.

Yu. T. Kalinin. Leutenant general, doctor of technical sciences, professor, head of Biopreparat from 1979 till the late nineties.

L. B. Kamenev (Rosenfeld) (1883–1936). Noted political man, revolutionist. He was against of the October coup (1917). After the revolution he took up a number of state posts. Along with Stalin and Zinov'v (see below) he fought against Trotsky (see). He was a participant of the "new" (1925–1927) and other oppositions (1925–1927) and was shot as a "enemy of nation" in 1936. Afterwards he was rehabilitated.

V. Kankalik. Professor and head of a university in Grozny (Chechnya). He was killed at the beginning of the rebellion in 1991. Just before his death, he organized a new medical faculty.

P. N. Kashkin. Outstanding expert in medical mycology.

I. A. Kassirskij. Noted scientist and clinical physician.

R. B. Khesin-Lurie. Noted molecular biologist, corresponding member of the Soviet Academy of Sciences.

N. S. Khrushchev (1894–1971). Outstanding leader of C.P.S.U., held a number of the high posts in the Soviet Union. The last of these were posts of the First Secretary of the Central Committee of C.P.S.U. (1953–1964) and prime minister (1958–1964). He uncrowned a cult of personality and activity of Stalin at congresses of the C.P.S.U. in 1956 and 1961 and began rehabilitation of victims of his repressions (Khrushchev's thaw). However, he retained the totalitarian regime in the country, suppressed the dissident movement, gave orders to shoot a demonstration of the working men in Novocherkassk (near Rostov) in 1962, and stifled a rebellion in Hungary in 1956, caused a crisis between the USSR and the USA (1962), and so on. He was displaced in 1964. His memoirs were published in New York in 1981.

L. N. Klassovskij. Expert in plague, doctor of medical sciences, professor of the Antiplague Institute in Alma-Ata (former capital of Kazakhstan).

L. A. Klyucharev. Major general, doctor of medical sciences, professor, deputy head of department of the Interagency Scientific and Technology Council on Molecular Biology and Genetics (1973–1979), then a head of department of science of Biopreparat. Before Moscow was at the Military Institute in Zagorsk (now Sergiev Posad, about 60 kilometers north of Moscow). He took part in the war with Germany (1941–1945).

W. Knapp. Well-known German expert in microbiology, the chairman of the International subcommittee on the taxonomy of pasteurellae, the genus of bacteria with which the plague microbe was a long time associated.

N. K. Koltsov (1872–1940). Biologist. He was the first who hypothesized molecular structure of genes and their matrix functions.

G. I. Kondrashev. Senior researcher fellow of my laboratory at the All-Union Institute for Biological Instrument.

A. I. Korotyaev. Noted microbiologist, doctor of medical sciences, professor of the Medical Institute in Krasnodar (in the North Caucasuas).

V. S. Koshcheev. Corresponding member of the Academy of Medical Sciences. A short time he was a head of the Third Main Directorate under the USSR Ministry of Health. He now lives in the United States.

A. N. Kosygin (1904–80). Prime minister of the Soviet Union (1964–80) and a member of the Politburo of the Central Committee of CPSU (1948–52 and 1960–80).

V. A. Kraminskij. Ex-colonel, noted expert in epidemiology, doctor of medical sciences, professor at the Irkutsk Antiplague Institute. He took part in the war with Germany.

A. S. Kriviskij. Expert in genetics of bacteria, doctor of biological sciences, professor at the Institute of Genetics of Industrial Microorganisms in Moscow.

Svetlana Lebedeva. Expert in genetics of plague microbe, doctor of medical sciences, professor. She works at the Rostov Antiplague Institute.

V. A. Lebedinskij. Lieutenant general, microbiologist and immunologist, academician of the Soviet Academy of Medical Sciences. He was a member of the Interagency Scientific and Technology Council on Molecular Biology and Genetics (1973–75). After E. Smirnov (see below), he was briefly a chief of the organization P.O. A-1968.

Galina Lenskaya. Expert in plague, candidate of medical sciences, ardent adherent of T. Lysenko and N. Zhukov-Verezhnikov (see below). She worked at the Saratov Mikrob Institute.

M. I. Levi. Noted expert in plague, doctor of medical sciences, professor. He worked at the Rostov Antiplague Institute in the late fifties and early sixties. After he was in Moscow till his death in 2002.

M. G. Lokhov. Expert in brucellosis. He worked at the Saratov Mikrob Institute.

Yu. M. Lomov. Expert in cholera, doctor of medical sciences, professor, director of the Rostov Antiplague Institute.

Ludmila D. Lukyanova. Corresponding member of the Soviet Academy of Medical Sciences, wife of the speaker of the last Supreme Soviet of the USSR.

A. M. Lunts. Biologist, professor of the Saratov Medical Institute, follower of classical genetics. In 1947, after session of the All-Soviet Union Agricultural Academy named after V. I. Lenin, he was a subject to repression as an opponent of T. Lysenko.

T. D. Lysenko (1898–1976). Agronomist, founder of pseudoscientific "Michurin doctrine" in biology ("lysenkoism"). In the basis of the doctrine there were two dogmatic assertions: Hereditary transfer of signs acquired during the life of individual (thus Darwinism was replaced by Lamarckism) and denial of existence of gene—special substance, responsible for genetic inheritance of signs (thus the chromosome inheritance theory was rejected). The opponents of Michurin's biology were accused of being idealists and metaphysicians, worshiping foreign ideas. Their studies and research were considered to be in conflict with the platform of the Party and the government. As a result, thousands of biologists, researchers, and teachers were expelled from their positions and replaced by ignorant or unprincipled people. The teaching and scientific books, containing materials contradicting Michurin's biology, were withdrawn from the libraries, and sometimes destroyed. Lysenko was supported by Stalin; for Stalin it was important that the science should be divided into his own, the only materialistic and progressive one, and bourgeois, outmoded pseudoscience. In his turn, Lysenko became Stalin's active assistant.

V. F. Mazovka. The second secretary of the Rostov Committee of CPSU.

G. M. Medinskij. Ex-lieutenant colonel, noted expert in epidemiology and field of civil defense, doctor of medical sciences, professor at the Rostov Antiplague Institute. He took part in the war with Germany (1941–45).

L. A. Melnikov. Microbiologist, candidate of medical sciences. He worked at the Saratov Mikrob Institute in the early fifties. Ex-colonel of KGB.

I. V. Michurin (1855–1935). Biologist-selectionist, creator of many new varieties of fruits and berries.

H. Mollaret. Well-known French expert in plague and other yersiniae, the secretary of the International subcommittee on the taxonomy of pasteurellae (see above, W. Knapp).

V. M. Molotov (Skryabin) (1890–1986). One of the outstanding leaders of CPSU and the Soviet Union. As the Minister of Foreign Affairs signed with fascist Germany the nonaggression pact and the secret agreement about the division of Europe in 1939. He was one of adherents of mass repressions in the USSR (he even did not object to arrest of own wife). He came out as an attorney of Stalin in the early fifties. After that he was removed from all leading positions and sent as an ambassador in Mongolia.

A. V. Naumov. Biochemist, doctor of medical sciences, professor, ex-director of the Saratov Mikrob Institute.

N. V. Nekipelov. Zoologist, expert in plague, doctor of biological sciences, professor at the Irkutsk Antiplague Institute.

Charles Nicolle (1866–1936). French bacteriologist, a Nobel Prize winner (1928). He discovered the role of louses as carriers of spotted fever agent.

N. L. Nikolaev. Major general, doctor of medical sciences, professor, director of Saratov Mikrob Institute in the sixties and in the early seventies. Before he was a head of the Military Institute in Kirov (now Vyatka, northeast from Moscow).

V. I. Ogarkov (1926–1987). Major general, microbiologist, corresponding member of the Academy of Medical Sciences, the first head of Biopreparat. Before his appearance in Moscow in 1973, he was a chief of the Military Institute in Sverdlovsk (Ekaterinburg, in the Ural).

Elisaveta Ordynskaya. Governess of my mother.

Yu. A. Ovchinnikov (1934–1988). Molecular biologist, academician of the Academy of Sciences and its vice president, director of the M. M. Shemyakin Institute of Bioorganic Chemistry, one of initiators of a new BW program of the USSR, a member of the Interagency Scientific and Technology Council on Molecular Biology and Genetics.

I. M. Papovyan. Surgeon, senior lecturer, from 1947 was a rector of the Saratov Medical Institute for a few years. In 1947 he was sent from the Ministry of Heath as its plenipotentiary for battle with opponents of T. Lysenko in Saratov.

V. A. Pasechnik. Chemist, candidate of chemical sciences, head of the Institute for Ultra-Pure Preparations in Leningrad (St. Petersburg). In 1989, while on a business trip to Paris, he refused to return home. After that began the collapse of the "Biopreparat" system.

B. N. Pastukhov. Head of department of dangerous infections of the Soviet Ministry of Health in the fifties.

V. N. Pautov. Microbiologist and epidemiologist, corresponding member of the Soviet/Russian Academy of Medical Sciences. He was a chief of the Military Institute in Kirov (Vyatka) in the late seventies and early eighties.

R. V. Petrov. Immunologist, academician of the Academy of Medical Sciences and the Academy of Sciences and its vice president (after Yu. Ovchinnikov's death). He was a member of the Interagency Scientific and Technology Council on Molecular Biology and Genetics.

V. S. Petrov. Zoologist, expert in plague, doctor of medical sciences, professor. He worked at the Antiplague Institute in Alma-Ata (Kazakhstan) in the fifties.

B. V. Petrovskij. Well-known surgeon, academician, from 1965 for twenty years was a Minister of Health.

V. A. Pertsik. Noted lawyer, doctor of law sciences, professor. He took part in composition of the last Soviet constitution in the seventies.

A. F. Pinigin. Expert in brucellosis, candidate of medical sciences, for a long time worked at the Irkutsk Antiplague Institute. He was the first who discovered brucellosis of reindeers.

V. I. Pokrovskij. Infectionist and epidemiologist; from the late eighties has been the president of the Academy of Medical Sciences.

A. M. Prokhorov. Physicist, academician of the Academy of Sciences, one of the creators of maser, Nobel Prize winner (1964, along with N. Basov and Ch. Townes).

A. A. Prozorov. Molecular geneticist, doctor of biological sciences, professor.

S. V. Prozorovskij. Microbiologist, academician of the Academy of Medical Sciences, a head of the N. F. Gamaleya Institute of Epidemiology and Microbiology in Moscow from the early eighties till the late nineties.

V. L. Pustovalov. Immunochemist, candidate of medical sciences. He worked at the Rostov Antiplague Institute.

S. M. Rassudov. Microbiologist, doctor of medical sciences, professor. He was a deputy director of the Rostov Antiplague Institute for a short time. He was also a director of Saratov Mikrob Institute in 1972–74.

I. I. Rogozin. Major general, professor at the Military Medical Academy in Leningrad (St. Petersburg), corresponding member of the Academy of Medical Sciences, expert in epidemiologist.

Sofiya Rotenberg. My mother's mother.

V. I. Rubin. Biochemist, doctor of medical sciences, professor of the Saratov Medical Institute.

G. P. Rudnev. Well-known expert in plague and other dangerous infections. His monograph *The Clinical Aspects of Plague* is still of great value.

I. V. Ruzhentsova. Laboratory assistant of my laboratory at the All-Union Institute for Biological Instrumentation.

L. A. Ryapis. Expert in field of dangerous infection, doctor of medical sciences, professor of the Moscow I. Sechenov Medical Academy.

V. N. Rybchin. Noted molecular geneticist.

R. S. Rychkov. For a short time was a head one of sections at the Central Committee of CPSU and then a head of Glavmikrobioprom and the Interagency Scientific and Technology Council on Molecular Biology and Genetics.

A. D. Sakharov (1921–1989). Soviet physicist and public figure, one of creators of the H-bomb in the USSR, academician of the Academy of Sciences. He began to come out against conducting a trial of nuclear weapons from the late fifties. He was one of the leaders of the dissident movement in the Soviet Union in the early seventies. He originated a project of new constitution of the USSR (1990). Nobel Peace Prize (1975).

V. D. Savve. Ex-colonel, doctor of biological sciences, professor, the first director of Institute of Immunology of Biopreparat in Lubuchany near Chekhov, a town about 60 kilometers south of Moscow. Formerly he worked at the Military Institute in Zagorsk (now Sergiev Posad, about 60 kilometers north of Moscow).

I. A. Sedin. Husband of Mariya, aunt of my mother, who was arrested by the GPU together with me father and other relatives.

Irina Semina. Research fellow of my laboratory at the All-Union Institute for Biological Instrumentation.

A. C. Serebrovsky (1892–1948). Biologist, antipode of Lysenko, one of the founders of home genetics. He confirmed experimentally the idea of gene divisibility. He was a corresponding member of the Academy of Sciences.

I. Shemyakin. Research fellow of the Institute of Applied Microbiology at Obolensk (near Moscow).

M. Sholokhov (1905–1984). Soviet writer (author of *Tikhy Don*), the Nobel Prize winner (1965).

G. M. Shub. Microbiologist, doctor of medical sciences, professor of the Saratov Medical Institute.

L. S. Shvarts. Therapeutist, doctor of medical sciences, professor of the Saratov Medical Institute. He suffered during "Doctors' Plot" events.

Vera Simagina. Therapeutist, doctor of medical sciences, professor, headed the subfaculty of diagnostics at the Saratov Medical Institute.

V. A. Sizov. Colonel, expert in anthrax, an official of Biopreparat.

Adelina Skavronskaya (1921–1999). Noted expert in genetics of bacteria, academician of the Soviet/Russian Academy of Sciences, head of department of genetics at the N. F. Gamaleya Institute of Epidemiology and Microbiology in Moscow, follower of V. Timakov (see below).

A. A. Skladnev. Ex-colonel. Chief one of departments of Biopreparat and member of the Interagency Scientific and Technology Council on Molecular Biology and Genetics from 1973 till 1975–77.

A. M. Skorodumov. Noted expert in plague, professor, founder of the Antiplague Institute in Irkutsk (1934). As with many other members of the plague control establishment, he met a tragic destiny. He was arrested in 1937 and perished in the dungeons of the NKVD.

G. K. Skryabin (1917–1988). Noted microbiologist and biochemist, academician of the Soviet Academy of Sciences and its Main Scientific Secretary, director of the Institute of Physiology and Biochemistry of Bacteria in Pushchino-on-Oka (about 120 kilometers south of Moscow), member of the Interagency Scientific and Technology Council on Molecular Biology and Genetics.

E. I. Smirnov (1904–1985). Colonel general, academician of the Soviet Academy of Medical Sciences. In 1947–53 was a Minister of Health. Then served at the Ministry of Health. Since 1960 was a head of the organization P.O. Box A-1968 (or the 15th Directorate of the Ministry of Defense), and from 1975 he was a member of the Interagency Scientific and Technology Council on Molecular Biology and Genetics. He took part in the war with Germany (1941–1945) as the Main Sanitary Inspector of the Ministry of Defense. Author of the book *War and Military Medicine* (Moscow, 1979).

L. V. Smirnov. Vice-premier minister of the USSR, head of the Military Industrial Commission.

V. P. Smirnov (?–1976). Remarkable expert in plague. He was inspired with the idea of vaccinating humans against plague through conjunctiva. While working in Mongolia, where he vaccinated more than one hundred thousand people, V. P. Smirnov infected himself with plague in order to demonstrate the effectiveness of his method. From the late fifties till his death in 1976 worked at the Irkutsk Antiplague Institute.

B. N. Sokov. Headed one of laboratories for tularemia and other infections study at the Institute of Applied Microbiology at Obolensk (near Moscow).

Lina Stern (1878–1968). Noted physiologist, academician of the Academy of Sciences and the Academy of Medical Sciences. She suffered during "Doctors' Plot" events.

Yu. G. Suchkov. Expert in plague, doctor of medical sciences, professor. He worked at the Rostov Antiplague Institute, then was a director of the Antiplague Institute in Stavropol (in the North Caucasus). He was moved to Moscow in 1983 as a deputy chief of the Third Main Directorate under the Ministry of Health.

Ya. M. Sverdlov (1885–1919). Noted statesman, a participant of the revolution in the Ural in 1905–1907 and of the October coup in 1917. He was a

Chairman of the All-Union Central Executive Committee and a Secretary of the Central Committee of Communist Party from 1917 till his death in 1919.

V. S. Tarumov. Colonel, expert in biological warfare technology, one of deputy directors of the Institute of Applied Microbiology in Obolensk (near Moscow). Before Obolensk, he worked at the Military Institute in Sverdlovsk in the Ural (now Ekaterinburg).

T. I. Tikhonenko. Noted expert in virus biochemistry, doctor of biological sciences, professor, deputy of V. Zhdanov (see below).

V. D. Timakov (1905–1977). Well-known microbiologist and public man; president of the Academy of Medical Sciences of the USSR (1968–77). He fought against "lysenkovshina."

Lidiya Timofeeva. Microbiologist, doctor of medical sciences, worked at the Irkutsk Antiplague Institute.

I. S. Tinker. Expert in plague, doctor of medical sciences, professor, deputy Director of the Rostov Antiplague Institute in the early fifties, a victim of "Doctors' Plot" events.

M. T. Titenko. Ex-colonel, candidate of medical science, deputy director of the Rostov Antiplague Institute. Before Rostov worked at the Military Institute in Kirov (now Vyatka, northeast of Moscow).

L. D. Trotsky (Bronstein) (1879–1940). One of the leaders and the ideologists of the Bolshevism, author of doctrine about "Permanent revolution," one of the organizers of the October coup (1917) and of the Red Army (1918), ardent adherent of mass repression, opponent of Stalin. He was arrested in 1927 and exiled abroad (1929). He was killed by order of Stalin in Mexico by an agent of NKVD. He wrote a book, *My Life* (1940).

V. M. Tumanskij. Noted expert in plague, doctor of medical sciences, professor of the Saratov Mikrob Institute.

V. A. Tutelyan. Expert in field of nutrition, corresponding member of the Academy of Medical Sciences, deputy director of the Institute of Nutrition in Moscow.

V. Umnov. Journalist, who was familiar with the subject of biological weapons in the Soviet Union.

N. N. Urakov. Major general, doctor of medical sciences, professor, chief of the Obolensk Institute of Applied Microbiology from 1982. Before Obolensk he worked at the Military Institute in Kirov (Vyatka, northeast of Moscow) as a deputy chief.

V. I. Vernadskij (1863–1945). Well-known Russian naturalist and philosopher, author of study about biosphere and its evolution into noosphere.

V. N. Vinogradov. Noted therapeutist, academician of the Academy of Medical Sciences, a victim of "Doctors' Plot" events.

V. D. Vinogradov-Volzhinskij. Major general, parasitologist, the first director of the Institute of Applied Microbiology in Obolensk (near Moscow). In the fifties worked at the Military Institute in Zagorsk (Sergiev Posad).

Faina Vogel. Sister of my mother's mother.

M. N. Volgarev. Sanitarian, academician of the Academy of Medical Sciences, head of the Institute of Nutrition of the Academy of Medical Sciences.

O. V. Volkov (1900–1996). Noted writer and active fighter for democracy in the Soviet Union, a victim of Stalin's terror.

K. I. Volkovoi. Colonel, expert in plague, doctor of medical sciences, professor at the Obolensk Institute of Applied Microbiology. Before worked at institutes of the Ministry of Defense.

A. A. Vorobev. Major general, microbiologist and immunologist, academician of the Soviet/ Russian Academy of Sciences, deputy chief of "Biopreparat" from 1979 till 1987. Before Moscow he worked at the Military Institute at Zagorsk (now Sergiev Posad, about 60 kilometers north of Moscow).

K. Voroshilov (1881–1969). Marshal of the Soviet Union. He was a People's Commissar (Minister) of Defense (1934–39). He was intimate with Stalin many long years.

M. S. Vovsi. Physician, academician of the Academy of Medical Sciences. He was a main physican of the Soviet Army 1941–50, one of the victims of "Doctors' Plot" events.

E. A. Yagovkin. Expert in biochemistry of bacteria, doctor of medical sciences, deputy director of the Rostov Institute of Microbiology, Epidemiology and Parasitology.

D. K. Zabolotnij (1866–1929). Well-known microbiologist and epidemiologist, academician of Soviet Academy of Sciences, president of the Ukrainian Academy of Sciences (1928–29). He proved the identity of bubonic and primary pulmonary plague and established role of marmota as carrier of plague microbe. He was one the first scientific researchers to accept the Soviet regime unconditionally and may have been involved in the idea of developing biological weaponry.

L. N. Zajkov. Member of the Political Bureau of Central Committee of CPSU in the eighties.

V. Zakrutkin. Soviet writer.

A. A. Zhdanov (1896–1948). Secretary of Central Committee (1934–48) and the first secretary of the Leningrad Committee of CPSU (1934–44). He was intimate with Stalin. One of the active organizers of mass terror in the USSR in the thirties and the forties.

V. M. Zhdanov (1914–1987). Noted expert in virology, one of the initiators of smallpox eradication, academician of the Soviet Academy of Medical Sciences and till his death (1987) the director of D. I. Ivanovsky Institute of Virology in Moscow. The first chief of the Interagency Scientific and Technology Council on Molecular Biology and Genetics (1973–75). He was a deputy minister of health in the late fifties.

I. F. Zhovtyj. Parasitologist, expert in plague, doctor of biological sciences, professor. For many years he was a deputy director of the Irkutsk Antiplague Institute.

N. N. Zhukov-Verezhnikov (1908–1981). Expert in the field of dangerous infections, academician of the Academy of Medical Sciences and its vice president 1949–52, Deputy Minister of health (1952–54), ardent adherent of the Soviet regime and T. Lysenko. He was the accuser during the Khabarovsk trial at Japanese war criminals (1949) and secretary of International Commission to investigate the supposed facts of bacteriological warfare by the U.S. Army in China and Korea in 1952.

L. A. Zilber (1894–1966). Virologist and immunologist, academician of the Academy of Medical Sciences. One of the first who hypothesized the role of viruses in tumor origin and began to study tumor immunology. One of the victims of Stalin's repression.

G. E. Zinov'v (Radomyslensky) (1883–1936). Prominent political man, a participant of Russian revolution in 1905–1907, an adversary of the October coup in 1917, president of the Executive Committee of the Comintern (1919–26), one of the opponents of L. Trotsky (see above) and Stalin. He was shot in 1936.

S. Zinovev. Chief of my grandfather.

BIBLIOGRAPHY

Aleksandrov, N. I., and Nina Gefen. *Aktivnaya spetsificheskaya profilaktika infektsionnykh zabolevaniy i puti eyo usovershenstvovaniya* (Active Specific Prophylaxis of the Infectious Diseases and Ways of Prophylaxis's Improvement). Moscow: Ministry of Defense Publishing House, 1962).

Barry, J., with Carroll Bogert et al. "Planning a Plague?" *Newsweek,* February 1, 1993, pp. 20–22.

Doklad mezhdunarodnoiy nauchnoiy komissii po rassledovaniyu faktov bakteriologicheskoiy voiyny v Koree i Kitae (Report of the International Scientific Commission to Investigate the Facts of Bacteriological Warfare in Korea and China). Beijing, 1952.

Domaradskij, I. V. *Chuma: sostoyanie, gipotesy, perspectivy* (Plague: Situation, Hypotheses, Prospects). Saratov: Saratov Medical Institute Publishing, 1993.

———. "Na pozhiznennyiy srok?" (A Life Sentence?). *Pravozachitnik* (Human Rights Defender), no. 4 (1995): 81–83.

———. "Istoriya odnoiy avantury" (The Tale of an Adventure). *Znanie-sila,* no. 11 (1996): 60–72, and no. 12 (1996): 54–64.

———, E. P. Golubinskij, S. A. Lebedeva, and Yu. G. Suchkov. *Biokhimiya i genetika chumnogo mikroba* (The Biochemistry and Genetics of the Plague Microbe). Moscow: Meditsina, 1974.

Lebedinskiy, V. A. *Ingalyatsionnyiy (aerogennyiy) metod vaktsinatsii* (The Inhalant [Aerogebic] Method of Vaccination). Moscow: Meditsina, 1971.

Leitenberg, M. "Biological Weapons: Scientist and Citizen." Special Issue. *Chemical and Biological Warfare* 9, no. 7 (1967): 153–67.

Leskov, S. *Chuma i Bomba* (Plague and a Bomb). *Izvestiya,* no. 118, June 26, 1993.

Levi, M. I., editor and compiler. *Zanimatel nye ocherki o deyatelnosti i deyatelyakh protivochumnoiy sistemy Rossii i Sovetskogo Soyusa* (A Few Entertaining Sketches of the Actors and Activities of the Plague-Control System in Russia and the Soviet Union). Moscow: Informatika, 1994, pt. 1, pp. 65–70, 158–64.

Medinskiy, G. M., M. I. Narkevich, Yu. M. Lomov, et al. *Spravochnik-kadastr rasprostraneniya vibrionov eltor v poverkhnostnych vodoemakh i stochnykh vodakh na territorii Sovetskogo Soyuza v 7-yu pandemiyu kholery* (A Reference Survey of the Spread of Eltor Vibrions in Bodies of Surface Water and Effluents in the USSR during the Seventh Cholera Pandemic). Edited by Yu. M. Lomov. Rostov-on-Don: Antiplague Institute Publishing House, 1991.

Medinskiy, G. M., Yu. M. Lomov, G. I. Kulov, et al. "Kolera" (Materialy Rossiyskoiy nauchnoiy konferents (Cholera: Proceedings of a Russian Scientific Conference). Rostov-on-Don, November 18–19, 1992. Rostov-on-Don: Antiplague Institute Publishing House, 1992.

Ogarkov, V. I., and K. G. Gapochko. *Aerogennye infektsii* (Aerogenic Infections). Moscow: Meditsina, 1975.

Rosebury, T. *Peace or Pestilence: Biological Warfare and How to Avoid It.* New York: Whittlesey House, 1949.

Rozhnyatovskiy, T., and Z. Zholtovskiy. *Biologicheskaya voiyna* (Biological Warfare). Moscow: Foreign Literature Publishers, 1959.

Rubinshteiyn, M. I. *Burzhuasnaya nauka i tekhnika na sluzhbe amerikanskogo Imperializma* (Bourgeois Science and Technology in the Service of American Imperialism). Moscow: USSR Academy of Sciences, 1951.

Smirnov, E. I. *Voyna i voennaya meditsina* (War and Military Medicine). Moscow, 1979.

Stockholm International Peace Research Institute. *The Problem of Chemical and Biological Warfare.* Stockholm: Almqvist & Wiksell, 1971, 1:287 and 5:238–58.

Umnov, V. "Posle 20 let molchaniya sovetskie mikroby zagovorili" (After the Twenty Years of Silence Soviet Microbes Begin to Speak). *Komsomolskaya pravda,* no. 80, April 30, 1992.

Velikanov, V. I. *Sudby liudskie* (People's Fates). Moscow, 1998.

Yankulin, V. I. "Sindrom chumy ili khozhdenie po mukam odnogo iz sozdateley bakteriologicheskogo oruzhiya" (Plague Syndrome or the Pass of Torments of a Bacteriological Weapon's Creator). *Izvestiya,* no. 196, October 15, 1997.

INDEX

Institute for Protein Biosynthesis, 251
Interagency Scientific and Technical
Council for Molecular Biology and
Genetics, 157
relations with Igor Domaradskij,
119–20, 122, 166, 198
relations with S. I. Alikhanyan, 164,
165
relations with V. I. Ogarkov, 193
Belyakov, V. D., 254
Berdnikov, A. I., 56–57
Beria, Lavrenty P., 14, 15, 102, 265,
307
Berlin, Abram Lvovich, 59, 128, 173,
186–87
binary preparations concept, 219–20
*Biochemistry and the Genetics of the
Plague Agent*, 199
biodefense program. *See* Problem No.
5
bioengineering, 174, 176
Biohazard, 9, 137, 305
See also Alibekov, Kanatjan
biological defense. *See* Problem No. 5
Biological Toxin and Weapons Convention, 1972, 18, 111, 158
hiding illegal research from, 118,
146, 151, 184–85
potential violations of, 245, 281
Biopreparat, 10, 199, 301
civilian/military friction, 118, 154,
183, 198, 204, 225
conversion from biological weapons
to vaccines, 277
funding, 169, 171, 278
hiring practices, 184, 200
immunological institutes, 163
institutes and facilities, 302
and new Obolensk lab, 234–35
Problem Ferment, 16

purpose to hide bioweapons research,
14, 118, 146, 151, 253, 274
reasons for Soviet failures, 287–88
revealed to the West, 14, 206, 211,
272–74
scientific goals, 147–48, 201
secrecy requirements, 152, 180–81,
200–201, 275
and tularemia research, 201,
202–203, 204
bioweapons
accidents, 103
against animals, 157–58
Bonfire plan, 177–78
British use of, 133
choice of, 158–59
danger of theft, 279–80
delivery system for, 126–28
German use of, 127, 133
historic references, 123–25
humane, 206
Japanese use of, 124–25, 131–32
military requirements, 218
reasons for Soviet failures, 247–48
research by prisoners, 130
tactical vs. strategic, 147
US use of, 124, 132, 147, 158, 210
VX release, 136
Warsaw Pact allies use of, 133–34
See also Problem Factor
bioweapons vs. nuclear weapons,
134–35
bivalent vaccine, 83
Black Death. *See* plague
Blokhina, Irina, 295
Bogert, Caroll, 276
Bogomolets, A. A., 56
Bonfire, 177–78
Borovik, R. V., 186, 199, 201
meeting on Domaradskij/Urakov rela-
tions, 242

Vinogradov-Volzhinskij, V. D.,
197–99, 201, 203, 218
dismissal from Institute of Applied
Microbiology, Obolensk, 213–14
Virology Institute, Sergiev Posad, 16,
302
virus theory of cancer, 130
Volgarev, M. N., 283
Volgograd Antiplague Institute. *See*
Antiplague Institute, Volgograd
Volkov, O. V., 265
Volkovoi, K. I., 200–201
Vorobev, A. A., 203, 204
Academy of Medical Sciences,
190–91, 235–36, 288
relations with Igor Domaradskij,
191–92, 194
theory to alter antigenicity, 204–205
Voroshilov, K., 126
Vovsi, M. S., 66
Vozrozhdeniye Ostrov. *See* Rebirth
Island
VPK. *See* Military Industrial Com-
mission

Waksman, Selman A., 58, 72
War and Military Medicine, 131, 316
weapons, tactical vs. strategic, 147
White Guards, 34
White House, Russian. *See* Russian
White House
WHO. *See* World Health Organization
World Health Organization, 85, 91
World War II and Igor Domaradskij,
46
Wu Lien-teh, 57

Yagovkin, E. A., 269
Yeltsin, Boris, 103, 135, 267, 268,
269, 271
decree on biological weapons, 278
Yersina pseudotuberculosis, 182, 192
Yersinia pestis, 85, 110, 116, 124
EV strain, 115
plasmids, 167, 182
Young Communists, 50–51

Zabolotnij, D. K., 56, 76, 125–26
Zajkov, Lev Nikolaevich, 237, 238,
247
Zakrutkin, V., 106
Zavyalov theory of antigenicity, 205
Zdrodovskij, P. F., 129
Zhdanov, A. A., 106
Zhdanov, V. M., 13, 83, 142–43, 149,
317
graft, 155–56
Interagency Scientific and Technical
Council for Molecular Biology and
Genetics, 145, 157
Zhdanov, Yu., 106
Zhovtyj, I. F., 60, 79, 85, 91
Zhukov-Verezhnikov, Nikolay N., 97,
311
cholera treatment, 99
investigations of biological war
crimes, 131, 132
lysenkovshchina follower, 65
pneumonic plague treatment, 58
Zilber, L. A., 130, 308
Zilinskas, Raymond, 169, 171
Zinovev, S., 38
Zinov'v (Radomyslensky), G. E., 310